SpringerBriefs in Computer Science

For further volumes:
http://www.springer.com/series/10028

Wei Wang • Qian Zhang

Location Privacy Preservation in Cognitive Radio Networks

 Springer

Wei Wang
Department of Computer Science
and Engineering
Hong Kong University of Science
and Technology
Kowloon, Hong Kong SAR

Qian Zhang
Department of Computer Science
and Engineering
Hong Kong University of Science
and Technology
Kowloon, Hong Kong SAR

ISSN 2191-5768   ISSN 2191-5776 (electronic)
ISBN 978-3-319-01942-0   ISBN 978-3-319-01943-7 (eBook)
DOI 10.1007/978-3-319-01943-7
Springer Cham Heidelberg New York Dordrecht London

Library of Congress Control Number: 2013949147

Springer is part of Springer Science+Business Media (www.springer.com)

Preface

Cognitive radio networks (CRNs) have been considered as an attractive means to mitigate the spectrum scarcity issue that is expected to occur due to the increasing demand for wireless channel resources. In CRNs, unlicensed users can opportunistically access temporarily available licensed bands or white spaces according to spectrum availability information, which can be obtained by collaborative spectrum sensing or database query. As such, the unlicensed users need to share their location-related information with other unlicensed users or the database owner, which may compromise their location privacy. This book focuses on the current state-of-the-art research on location privacy preservation in CRNs. Along with the reviewed existing works, this book also includes fundamental privacy models, possible frameworks, useful performance, and future research directions.

Kowloon, Hong Kong SAR

Wei Wang
Qian Zhang

Contents

Acronyms

CLT	Central limit theorem
CRN	Cognitive radio networks
DLC attack	Differential report-location correlation attack
DM	Discernibility metric
EGC	Equal gain combining
EMD	Earth mover distance
FC	Fusion center
FCC	Federal communications commission
GPS	Global positioning system
i.i.d.	Independent and identically distributed
LBA	Location-based advertising
LBS	Location-based service
LDC	Locally decodable codes
MD	Minimal distortion
p.d.f.	Probability density function
PIR	Private information retrieval
POI	Points of interest
pmf	Probability mass function
PU	Primary user
QI	Quasi-identifier
RLC attack	Report-location correlation attack
RSS	Received signal strength
SA	Sensitive attribute
SNR	Signal-to-noise ratio
SINR	Signal-to-interference-plus-noise ratio
SP	Service providers
SPRT	Sequential probability ratio test
SSE	Sum of squared error
SU	Secondary user
USRP	Universal software radio peripherals
WSD	White space device

Chapter 1
Introduction

The rapid proliferation of wireless technology offers the promise of many societal and individual benefits by enabling pervasive networking and communication via personal devices such as smartphones, PDAs, computers. This explosion of wireless devices and mobile data creates an ever-increasing demand for more radio spectrum. The spectrum scarcity issue is expected to occur due to the limited spectrum resources. However, previous studies [15] have shown that the usage of many spectrum bands (e.g., UHF bands) is inefficient, which motivates the concept of cognitive radio networks (CRNs) [22, 24, 25]. In CRNs, secondary (unlicensed) users (SUs) are allowed to access licensed spectrum bands given that it only incurs minimal tolerable or no interference to primary (licensed) users (PUs).

One of the major technical challenges in CRN is for SUs to exploit spectrum opportunity which refers to a time duration on a channel during which the channel can be used by SUs without interfering with the PUs operating on the channel. To achieve this goal, FCC (Federal Communications Commission) has recently adopted rules, including spectrum sensing and database query, to allow SUs to discover spectrum opportunities. SUs that are capable of spectrum sensing can sense spectrum bands to discover unused channels and the present of PUs. Among many spectrum sensing techniques, *collaborative spectrum sensing* has been considered as a promising way to ensure sufficient protection by exploiting sensor location diversity. Recent standard proposals for CRNs (e.g., IEEE 802.22 WRAN [3], CogNeA [2]) adopt collaborative sensing to improve spectrum sensing performance, that is, the sensing data from multiple SUs is aggregated to learn the spectrum occupancy. On the other hand, in a realistic CRN setting, multiple service providers (SPs) operate on the same set of frequency bands in one geographic area, where each SP serves a group of SUs. The latest FCC's rules [27, 28] have released the TV White Space for secondary access under a database-driven architecture, as depicted in Fig. 1.1 where there are several geo-location databases which are managed by the database operators and provide spectrum availability of TV channels. In database-driven CRNs, SUs are required to query a geo-location database for vacant channels at its location.

W. Wang and Q. Zhang, *Location Privacy Preservation in Cognitive Radio Networks*,
SpringerBriefs in Computer Science, DOI 10.1007/978-3-319-01943-7_1,
© The Author(s) 2014

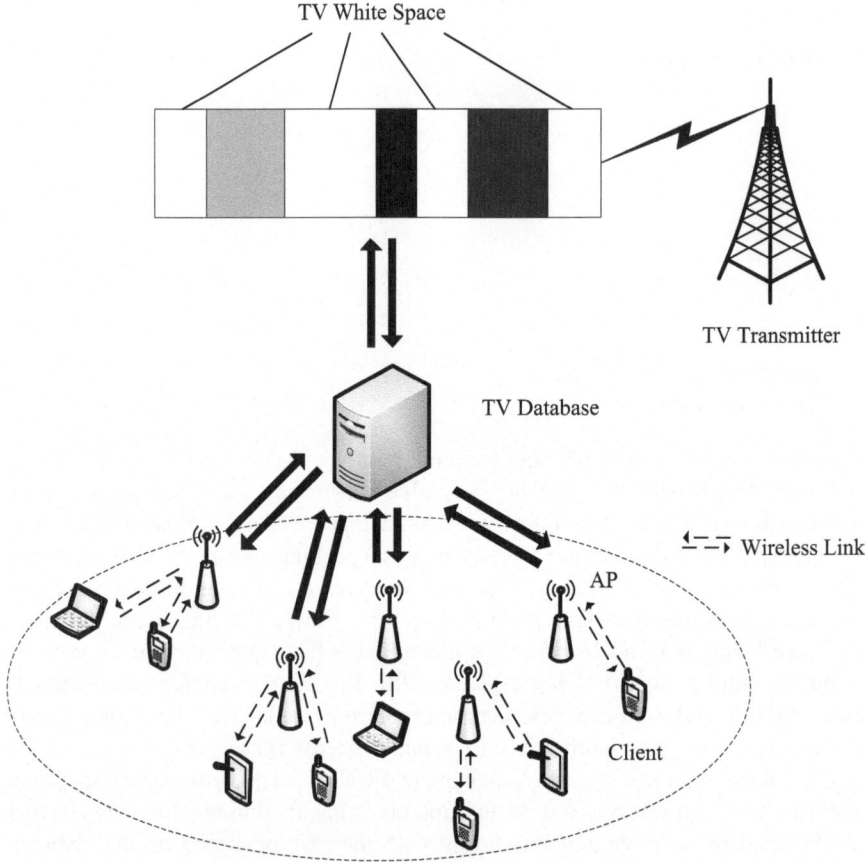

Fig. 1.1 Database-driven cognitive radio networks on TV white space

However, the open nature of wireless communication as well as software defined radio platforms, such as Universal Software Radio Peripherals (USRP) [56] and Sora software radio platform [91] makes CRNs face many new challenges in the aspects of location privacy. Nowadays, the growing privacy threats of sharing location information in wireless networks have been widely concerned [106]. The fine-grained location data can be used to determine a great deal about an individual's beliefs, preferences, and behavior. For example, an adversary may learn that a user regularly goes to a hospital by analyzing its location trace, which the user does not want to be disclosed, and sell such information to pharmaceutical advertisers. Moreover, malicious adversaries with criminal intent could hack the applications with such information to pose a threat to individual security and privacy. Being aware of such potential privacy risks, SUs may not want to share its data with fusion center or database, which, however, disables itself enjoying the benefits from collaborative spectrum sensing and database-driven CRNs if their privacy is not guaranteed. Therefore, it is essential to guarantee the privacy of each SU in collaborative sensing.

To enable SUs to enjoy services provided by CRNs, privacy preserving approaches are required. The main challenge of designing preserving privacy protocols for CRNs is to modify the data exchanged with other entities in CRNs so that others acquire no sensitive information while at the same time SUs can enjoy the CRN services without violating regulations. This raises four questions:

- What kind of sensing data or location data is safe for SUs?
- How to guarantee that no sensitive information is leaked?
- How to explore better tradeoffs between privacy and channel availability detection accuracy?
- How to provide privacy guarantee under the regulations?

We answer these questions briefly first. The reminder of this book will provide more details.

For the first question, there are plenty of privacy models to quantify privacy loss. Many partition-based privacy models are proposed to tackle different privacy concerns. The most basic and commonly used k-anonymity [86, 89] is developed to prevent adversaries from re-identifying an individual with a probability higher than $\frac{1}{k}$, where a value provided by a user is indistinguishable from those of $k - 1$ other users to provide privacy guarantee. A major axis considered in data publishing is differential privacy [34], whose idea is that the removal or addition of a single record in the dataset does not significantly affect the outcome of any data analysis. Entropy is also adopted to evaluate information leakage.

Consider the second question. Numerous techniques have been proposed for preserving privacy by modifying or transforming the original data. Basically, these techniques can be divided into four main categories: random perturbation, differential privacy, anonymization, and cryptographic methods.

First, random perturbation hides the real value of sensitive data by converting them into other values via a random process. In particular, random perturbation transforms the original data by replacing a subset of the data points with randomly selected values [6, 64, 65], or distorts the original data by adding random noise [5]. However, none of them can achieve the same individual privacy strength provided in this paper. To the best of our knowledge, the only work that studies the privacy issue in the context of collaborative sensing is by Li et al. [64]. Li et al. [64] identify a location privacy leakage problem in which an untrusted fusion center attempts to geo-locate a user by comparing the differences in aggregated results, and proposes a homomorphic encryption protocol and a distributed dummy report protocol to handle different types of attacks. In the distributed dummy report protocol, each user randomly makes a choice between sending the true report or sending a dummy report. However, this approach only considers a single service provider, while we study the scenario of multiple service providers. Moreover, the scheme proposed in [64] only measures the adversary's overall uncertainty on users' location, which cannot provide the differential privacy guarantee (at individual level) as described in this paper.

Second, differential privacy has been considered as a major axis in data publishing. Publishing different types of data is studied, such as histogram [12,111], set-valued data [17] and decision trees [38]. Among these studies, the data type related to our work is histogram. Blum et al. [12] divides the input counts into bins of roughly the same count to construct a one-dimensional histogram. By observing that the accuracy of a differential privacy compliant histogram depends heavily on its structure, Xu et al. [111] propose two algorithms with different priorities for information loss and noise scales. However, all these works discuss general data publishing, and assume one trustworthy server to process data, which is quite different from the problem considered in this paper.

Third, spatial cloaking and anonymization are widely adopted to preserve privacy in location-based services and participatory sensing [42,47,88,102], where a value provided by a user is indistinguishable from those of $k - 1$ other users to provide privacy guarantee, known as k-anonymity. Gedik and Liu [42] devises a framework which provides k-anonymity with different context-sensitive personalized privacy requirements. Several clique-cloak algorithms are proposed in [42] to implement the framework by constructing a constraint graph. In [102], locality-sensitive hashing is utilized to partition user locations into groups that contain at least k users. A form of generalization based on the division of a geographic area is adopted by *Anonysense* [88], where a map of wireless LAN access points is partitioned. *KIPDA* [47] enables k-anonymity for data aggregation with a maximum or minimum aggregation function in wireless sensor networks. Nevertheless, none of these works consider multiple service providers, or user collusion.

The fourth category preserves privacy via cryptographic techniques. Girao et al. [44] aggregate data based on homomorphic encryption, which preserves privacy by performing certain computations on ciphertext. The limitation of homomorphic encryption is that a server must know all the users that have reported data to compute the final aggregated results. Secure information aggregation frameworks are proposed in [81]. However, these works fail to provide privacy protection at individual level.

Let's consider the third and fourth questions. In CRNs, a fundamental task of each SU is to detect the presence and absence of PUs or to identify available spectrum bands at the SU's location. Modification on SU's sensing data or location data can preserve privacy, which, however, on the other hand, may compromise the accuracy of detecting channel availability. The modified data may compromise the detection accuracy of channel availability and lead to a wrong access decision, which causes interference with PUs. The interference not only costs throughput loss to SUs but also violate FCC's regulations. Moreover, in database-driven CRNs for TV White Space, FCC regulations require each enabling STA (e.g., Secondary Access Point) with localization ability of 50 m resolution. Thus, any modification on the location data should align with this regulation.

The above concerns make privacy preserving in CRNs more challenging than location privacy protection in general wireless and mobile networks. Several existing works [39, 40, 64] have proposed privacy preserving framework in the context of collaborative spectrum sensing and database-driven CRNs to prevent privacy leakage under the regulations of FCC. Li et al. [64] propose a homomorphic encryption protocol and a distributed dummy report protocol to handle different types of attacks. In the distributed dummy report protocol, each user randomly makes a choice between sending the true report or sending a dummy report while maintaining the usefulness of the aggregated results. Gao et al. [39, 40] leverage private information retrieval technique to thwart location privacy leaking from database query without modifying genuine location data.

In the reminder of this section, we will describe fundamental concepts of cognitive radio networks. We will also discuss some intuitions and observations about privacy issues in collaborative spectrum sensing and database-driven cognitive radio networks.

1.1 Cognitive Radio Networks on Licensed Bands

In this section, we will describe some fundamental concepts and key technologies of cognitive radio networks. We will first give a big picture of cognitive radio networks. To enable cognitive radio networks, SUs are usually required to be capable of spectrum sensing to detect the appearance of PUs. The state-of-the-art spectrum sensing techniques will be introduced. To cope with shadowing fading, cooperative spectrum sensing is proposed to enhance the sensing performance by exploiting the spatial diversity among SUs. On the other hand, FCC's rule in May 2012 [28] enforces database-driven CRNs and eliminates spectrum sensing as a requisite for cognitive radio devices.

1.1.1 Spectrum Sensing

In the spectrum sensing based CRNs, SUs can sense spectrum bands at their locations and dynamically access available spectrum [54]. The typical operations of an SU is shown in Fig. 1.2. To opportunistically access vacant channels, SUs need to sense the spectrum first [117]. And then based on the sensing results, SUs analyze the spectrum to find available sub-bands, and select the most appropriate sub-band with a proper transmission power without causing interference to PUs.

In terms of availability, spectrum sub-bands can be categorized into the following three classes:

- *White spaces* refer to the sub-bands that are not used by PUs at current time slots. Normally, white spaces are free of interference and only suffer from natural noises such as thermal noises.
- *Gray spaces* refer to the sub-bands that are under-utilized at current time slots. These sub-bands are usually occupied by limited interference from PUs as well as noises.

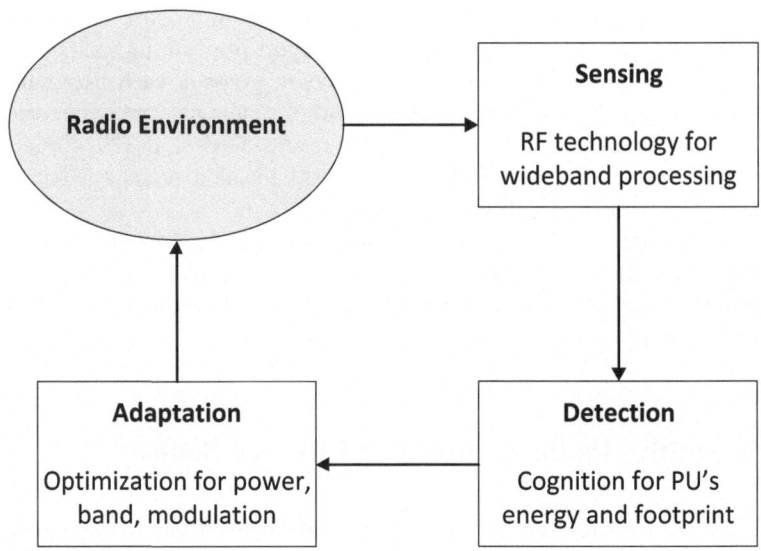

Fig. 1.2 Spectrum sensing cycle

- *Black spaces* are completely occupied by PUs. PUs will severely interfere with SUs if the SUs operate in black spaces.

Available spectrum bands can be made up of white spaces and gray spaces. The practical reality of re-using white spaces is quite straightforward since the SUs on the white spaces will cause little interference with PUs. While an essential observation on gray spaces is that a PU's signal may be too weak to be utilized in the local neighborhood, in which case the same sub-bands can be re-used by local SUs with low transmission power.

As the first step towards the above opportunities, spectrum sensing is a particular task on which the very essence of sensing-based cognitive radio networks rest. The goal of spectrum sensing is to find spectrum vacancies, i.e., the sub-bands of the spectrum that are non-utilized (white spaces) or under-utilized (gray spaces) at current time slots, by sensing the bands at the location of the SU in an unsupervised manner.

In spectrum sensing stage, there are normally two ways used to detect PUs to protect their service quality: physical-layer sensing and MAC-layer sensing, as illustrated in Fig. 1.3. Physical-layer sensing adapts SU's modulation schemes and other parameters to measure and detect the signals of PUs on a set of sub-bands, while MAC-layer sensing is used to determine when and which channel the SU should sense. The most popular sensing techniques include cyclostationary signal processing [16, 92] and radiometric detectors [87] (i.e., energy detectors) are generally used by SUs for physical-layer sensing. Cyclostationary signal processing detect PU according to an inherent property of digital modulated signals that naturally occur in the transmission of communication signals over a

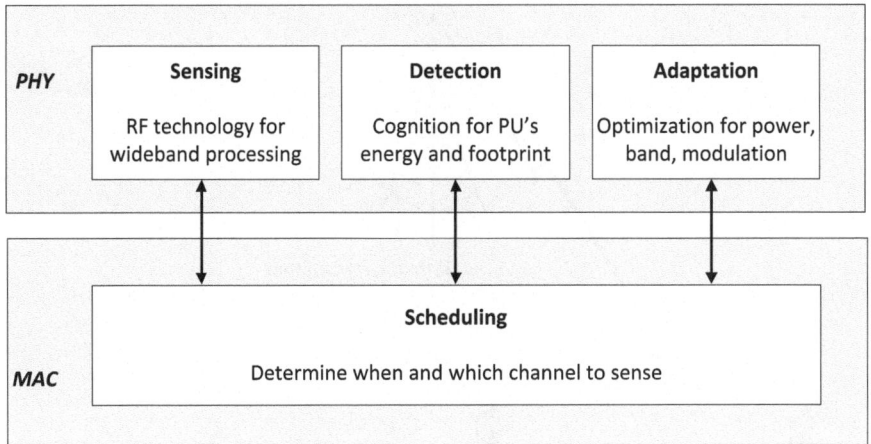

Fig. 1.3 Physical-layer sensing and MAC-layer sensing

wireless channel. The cyclostationarity approach to detection has an advantage over the energy-detection approach in that it is also capable of signal classification and has the ability to distinguish co-channel interference. However, the cyclostationarity approach is limited to PUs using standard modulation, while other types of PU (e.g., wireless microphone has no standard modulation specified by the FCC [26]) are usually detected by energy detection.

In both of these two sensing techniques, there are two models for channel availability from a secondary device perspective:

- *Binary model [24].* In binary model, a channel or sub-band is regarded as being occupied or non-utilized based on the presence or absence of any PU signal on that channel.
- *Interference temperature model [23].* In the interference temperature model, a channel or sub-band is considered as unavailable for transmission if the usage of the channel by an SU lead to interference temperature beyond a pre-defined threshold within its interference range.

In binary model, spectrum sensing at an SU can be modeled as a binary hypothesis-testing problem as follows.

$$\begin{cases} \Lambda_T \geq \eta_1 & \Rightarrow \text{accept } \mathscr{H}_1 \\ \Lambda_T \leq \eta_0 & \Rightarrow \text{accept } \mathscr{H}_0 \end{cases} \tag{1.1}$$

where Λ_T is the test statistics (e.g., energy) and η_1, η_0 are pre-defined thresholds. Specifically, hypothesis $\mathscr{H}_1$ refers to the presence of a PU's signal, i.e., the subband under test is occupied, while hypothesis $\mathscr{H}_0$ refers to the presence of ambient noise, i.e., the absence of PU's signal.

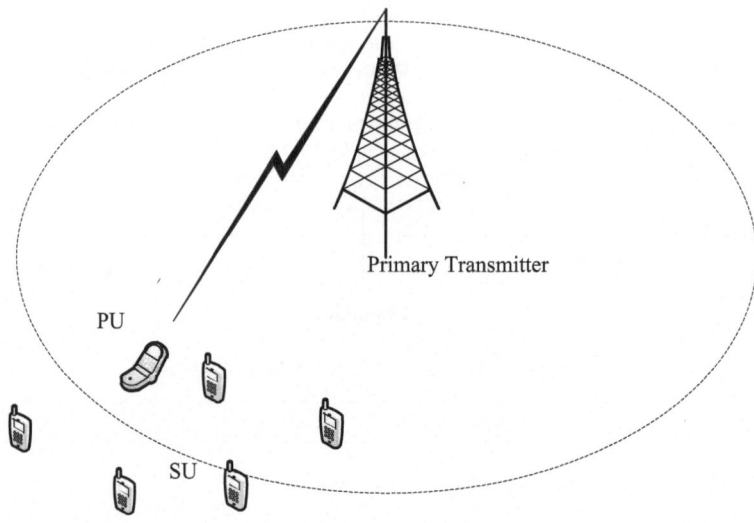

Fig. 1.4 The hidden terminal problem in CRNs

Interference temperature a measure of the power and bandwidth occupied by interference, and is normally defined as [21]:

$$T_I(f, B) = \frac{P_I(f, B)}{kB},$$ (1.2)

where $P_I(f, B)$ is the average interference power centered at frequency f, B the covering band, and k Boltzmann's constant.

In the interference temperature model, the FCC establishes an interference temperature limit $\hat{T}$ for a given geographic area. $\hat{T}$ stands for the maximum amount of tolerable interference for a given sub-band in a particular geographic area. Any SU's transmission must ensure its interference adding existing interference never exceeds $\hat{T}$.

Note that this book focuses on privacy issues in CRNs. For more detailed sensing techniques, please refer to [13, 50, 53].

1.1.2   Collaborative Spectrum Sensing

The key challenge in spectrum sensing is that performance of spectrum sensing degrades in shadowing or fading environments, where hidden terminal problem happens when the PU is shadowed. For example, as depicted in Fig. 1.4, when a PU is located within a high building, the presence of the PU may not be sensed

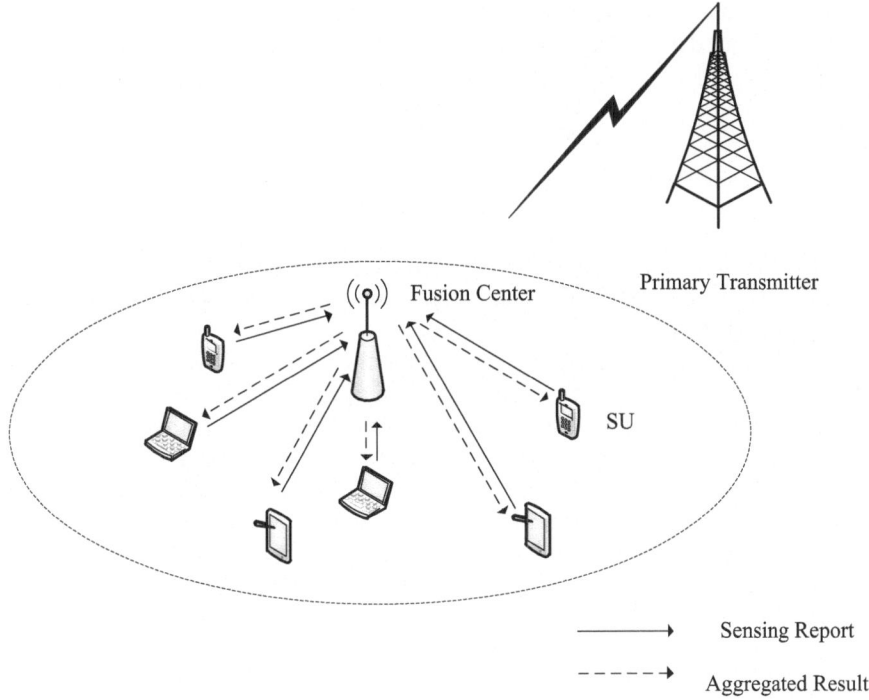

Fig. 1.5   Cooperative spectrum sensing in cognitive radio networks

by SUs outside the building due to the very low signal-to-noise ratio (SNR) of the received PU signal. As such, the observed channel is considered as vacant and can be accessed by the SUs, while the PU is still in operation and thus results in interference. To address this issue, multiple cognitive radios can be coordinated to performance spectrum sensing cooperatively. Several recent works have shown that cooperative spectrum sensing can greatly increase the probability of detection in fading channels. Recent standard proposals for CRNs (e.g., IEEE 802.22 WRAN [3], CogNeA [2]) adopt collaborative sensing to improve spectrum sensing performance, that is, the sensing data from multiple SUs is aggregated to learn the spectrum occupancy. For example, when the received signal of an SU located far away from a PU may be too weak to detect to the PU's signal, while by employing SUs nearby the primary user as collaborative detectors, the signal of the PU can be detected reliably.

The typical collaborative spectrum sensing is depicted in Fig. 1.5. For simplicity it is usually assumed that all SUs in the collaboration experience independent and identically distributed (i.i.d.) fading and shadowing with the same average

SNR. Under this assumption, the signal strength at sensor can be approximated as Gaussian using the Central Limit Theorem (CLT). Then, the distribution is given as [87]:

$$r_u \sim \begin{cases} \mathcal{N}\left(N_o, \frac{N_o^2}{M}\right) & \mathcal{H}_0 \\ \mathcal{N}\left(P_u + N_o, \frac{(P_u+N_o)^2}{M}\right) & \mathcal{H}_1 \end{cases} \tag{1.3}$$

where r_u is the u's RSS, N_o the noise power, and M the number of signal samples, e.g., 6×10^3/ms for 6 MHz TV band at the Nyquist rate. $\mathcal{H}_0$ stands for absence of PU signal and $\mathcal{H}_1$ for presence of PU's signal.

At each quiet period, a fusion center (FC) directs a set of SUs $\mathcal{S}$ to perform sensing for a certain length of sensing duration, e.g., 1 ms, and the SUs report their sensing results to the FC for data fusion at the end of each quiet period. The sensing result of an SU $S_i \in \mathcal{S}$ can be one bit to indicate the presence / absence of PU as a binary variable $T_i \in \{0 \text{ (the channel is idle)}, 1 \text{ (the channel is busy)}\}$, in which case the communication overhead is minimized while the sensing accuracy could be compromised. Over Rayleigh fading channels, the average probability of false alarm, the average probability of detection, and the average probability of missed detection are given by Digham et al. [31], respectively

$$P_f^{(i)} = \frac{\Gamma(\omega, \eta_i)}{\Gamma(\omega)},$$

$$P_d^{(i)} = \exp(-\eta_i/2) \sum_{p=0}^{\omega-2} \frac{1}{p!} \left(\frac{\eta_i}{2}\right)^p + \left(\frac{1+\gamma_i}{\gamma_i}\right)^{\omega-1}$$

$$\cdot \left[\exp(-\eta_i/(2(1+\gamma_i))) - \exp(-\eta_i/2) \sum_{p=0}^{\omega-2} \frac{1}{p!} \left(\frac{\eta_i \gamma_i}{2(1+\gamma_i)}\right)^p \right],$$

and

$$P_m^{(i)} = 1 - P_d^{(i)}. \tag{1.4}$$

Where ω denotes the time and width product, γ_i the average SNR at SU S_i, $\Gamma(a, x)$ the incomplete gamma function defined by $\Gamma(a, x) = \int_x^{\infty} t^{a-1} e^{-t} dt$ and $\Gamma(a)$ the gamma function, and η_i the energy detection threshold.

At the end of each quiet period, the FC applies certain fusion rule to process the collected sensing reports to make a final decision. The most commonly used fusion model for a single-round sensing is Equal Gain Combining (EGC), which is known to have near-optimal performance without requiring the estimation of channel gains [90]. Then the final decision statistic at the FC can be expressed as

$$T_\Sigma = \sum_{i:\{S_i \in \mathcal{S}\}} T_i, \tag{1.5}$$

Then, the FC makes a final decision based on the aggregated result T_Σ. A simple yet effective decision rule is called n-out-of-N rule [60], which is suitable for the case in which SUs report binary results. The n-out-of-N rule is described as

$$\begin{cases} T_\Sigma < n & \Rightarrow \text{accept } \mathcal{H}_0 \\ T_\Sigma \leq n & \Rightarrow \text{accept } \mathcal{H}_1, \end{cases} \tag{1.6}$$

where $1 \leq n \leq |\mathcal{S}|$. When $n = 1$, the n-out-of-N rule boils down into the OR-rule; when $n = |\mathcal{S}|$, the case of the n-out-of-N rule corresponds to the AND-rule. Previous studies [43,60] has shown that in many cases the OR-rule is the best among the fusion rules.

Based on the results given by (1.4), the false alarm probability of collaborative spectrum sensing based on the OR-rule is given by

$$P_f^\Sigma = 1 - \prod_{i=1}^{|\mathcal{S}|} \left(1 - P_f^{(i)}\right), \tag{1.7}$$

and the mis-detection rate of collaborative spectrum sensing based on the OR-rule is

$$P_m^\Sigma = \prod_{i=1}^{|\mathcal{S}|} P_m^{(i)}, \tag{1.8}$$

where $P_m^{(i)}$ is given by (1.4). We can see that for a given probability of false alarm, the probability of missed detection is greatly reduced when the number of collaborative SUs increases.

The sensing result can also be the exact signal strength statistics, i.e., T_i, which preserves more sensing information at the cost of more communication overhead. The exact signal strength statistics can be expressed as

$$T_i = h P_o \chi, \tag{1.9}$$

where h is the channel gain from the PU to the SU, P_o the transmission power of the PU, and $\chi \in \{0, 1\}$ the binary variable indicating the ON/OFF state of the PU. The channel gain h is usually characterized as

$$h = \left(\frac{d_o}{d_i}\right)^a e^{X_i} e^{Y_i}, \tag{1.10}$$

where d_o is the reference distance (e.g. 1 m), d_u the distance between the PU to the SU i, and a the path loss exponent. e^{X_i} and e^{Y_i} are shadowing fading and multi-path fading factors, respectively, where $X_i \sim \mathcal{N}(0, \sigma)$, $\forall i$. The multi-path fading is usually described as Rayleigh fading with zero mean, and thus $\mathbb{E}[e_i^Y] = 0$.

Typically, the decision rules for signal strength statistics report are based on hypothesis testing. One popular hypothesis testing technique is Walds Sequential

Probability Ratio Test (SPRT) [103]. SPRT is optimal in the sense of minimizing the average number of observations, given bounded false-alarm probability and mis-detection ratio. It enables the FC to reduce erroneous triggering of PU detection by optimizing its decision thresholds. The typical steps of SPRT are described as follows.

The vector of signal strength statistics collected by the FC in detection round t is denoted as $\boldsymbol{\theta}_t = [T_{1_t}, \ldots, T_{i_t}, \ldots, T_{N_t}]^{\mathrm{T}}$, where T_{i_t} is SU i's sensing report at time t. N_t is the number of collaborative SUs at time t. As shown in (1.3), signal strength statistics can be approximated as a Gaussian distribution in both presence (i.e., $\mathcal{H}_1$) and absence (i.e., $\mathcal{H}_0$) cases. Thus, decision rule problem is a binary decision problem to decide whether hypothesis $\mathcal{H}_0$ or $\mathcal{H}_1$ is true, given test statistic θ, where $\boldsymbol{\theta} = [\boldsymbol{\theta}_1^{\mathrm{T}}, \ldots, \boldsymbol{\theta}_t^{\mathrm{T}}, \ldots, \boldsymbol{\theta}_T^{\mathrm{T}}]^{\mathrm{T}}$.

To solve the binary decision problem, the FC keeps collecting new signal strength statistics from SUs until the amount of information and resulting testing performance being satisfied. To achieve this goal, SPRT is taken as the data processing rule to decide the stopping time and make a final decision. The main advantage of SPRT is that it requires fewer test statistics to achieve the same error probability, which is attained at the expense of additional computation. In the sequential decision process, the FC computes the log likelihood ratio and compares it with two thresholds η_0 and η_1. Either it determines on one of the two hypothesis, or it decides to take another round of statistic collection.

The likelihood ratio at detection round t is defined as:

$$\lambda_t = \ln \frac{p(\boldsymbol{\theta}_t | \mathcal{H}_1)}{p(\boldsymbol{\theta}_t | \mathcal{H}_0)}, \tag{1.11}$$

where $p(\boldsymbol{\theta}_t | \mathcal{H}_k)$ is the joint probability density function (p.d.f.) of test statistics collected at detection round t under hypothesis $\mathcal{H}_k$ $k \in 0, 1$. Recall that test statistics are assumed to be i.i.d. and follow Gaussian distribution, as shown in (1.3). Thus, (1.11) can be written as:

$$\lambda_t = \ln \frac{p(r_{1_t}, \ldots, r_{n_t} | \mathcal{H}_1)}{p(r_{1_t}, \ldots, r_{i_t} | \mathcal{H}_0)} = \sum_{i_t=1}^{n_t} \ln \frac{p(r_{i_t} | \mathcal{H}_1)}{p(r_{i_t} | \mathcal{H}_0)}, \tag{1.12}$$

where r_{i_t} is approximated as $r_{i_t} \sim \mathcal{N}(\mu_k, \sigma_k)$ under $\mathcal{H}_k$, according to Central Limit Theorem. Recall that $\sigma_0^2 = \frac{N_o^2}{M}$ and $\sigma_1^2 = \frac{P_u + N_o^2}{M}$, where P_u and N_o average noise power and received signal power at users. In a very low SNR environment, it is reasonable to approximate P_u as P_m, which is the minimal sensible signal. And thus approximate $(P_u + N_o)$ as N_o, and hence $\sigma_1 \approx \sigma_0$. Then, (1.12) can be expressed as:

$$\lambda_t = \frac{(\mu_1 - \mu_0) \sum_{i_t=1}^{n_t} r_{i_t} + \frac{1}{2} \sum_{i_t=1}^{n_t} (\mu_0^2 - \mu_1^2)^2}{\sigma_0}, \tag{1.13}$$

The next step is to determine the decision statistic Λ_T at detection round T is the joint likelihood ratio of a sequential test statistics $\boldsymbol{\theta}_1, \ldots, \boldsymbol{\theta}_T$ defined as:

$$\Lambda_T = \ln \frac{p(\boldsymbol{\theta}_1, \ldots, \boldsymbol{\theta}_T | \mathcal{H}_1)}{p(\boldsymbol{\theta}_1, \ldots, \boldsymbol{\theta}_T | \mathcal{H}_0)}, \tag{1.14}$$

where $p(\boldsymbol{\theta}_1, \ldots, \boldsymbol{\theta}_T | \mathcal{H}_k)$ is the joint p.d.f. of test statistics under $\mathcal{H}_k$). Regarding that the test statistics are Gaussian and i.i.d., we have:

$$\Lambda_T = \sum_{t=1}^{T} \ln \frac{p(\boldsymbol{\theta}_t | \mathcal{H}_1)}{p(\boldsymbol{\theta}_t | \mathcal{H}_0)} = \sum_{t=1}^{T} \lambda_t, \tag{1.15}$$

and based on (1.13) and (1.16), we finally have Λ_T as follows:

$$\Lambda_T = \frac{(\mu_1 - \mu_0)}{\sigma_0^2} \sum_{t=1}^{T} \sum_{i_t=1}^{n_t} r_{i_t} + \frac{1}{2\sigma_0^2} \sum_{t=1}^{T} \sum_{i_t=1}^{n_t} (\mu_0^2 - \mu_1^2). \tag{1.16}$$

The decision of SPRT at detection round T is based on the following rules [103]:

$$\begin{cases} \Lambda_T \geq \eta_1 & \Rightarrow \text{accept } \mathcal{H}_1 \\ \Lambda_T \leq \eta_0 & \Rightarrow \text{accept } \mathcal{H}_0 \\ \eta_0 < \Lambda_T < \eta_1 & \Rightarrow \text{take another detection round,} \end{cases} \tag{1.17}$$

where η_1 and η_0 are the detection thresholds, which are determined by predefined values of desired mis-detection rate α and false alarm rate β. However, the outage detection problem is opposite to detection problem described in [103] in sense of mis-detection rate and false alarm rate, since $\mathcal{H}_0$ is hypothesis for outage occurrence while $\mathcal{H}_1$ for event occurrence in [103]. Thus, the relationships in decision rule problem should be as follows:

$$\eta_1 = \ln \frac{1 - \alpha}{\beta},$$

$$\text{and } \eta_0 = \ln \frac{\alpha}{1 - \beta}, \tag{1.18}$$

while actual false alarm rate and mis-detection rate could be higher than α and β, respectively [101].

1.1.3 Database-Driven Cognitive Radio Networks

Besides spectrum sensing, database query is another typical approach to identify locally re-usable channels for the SUs. The latest FCC's rule [28] released in

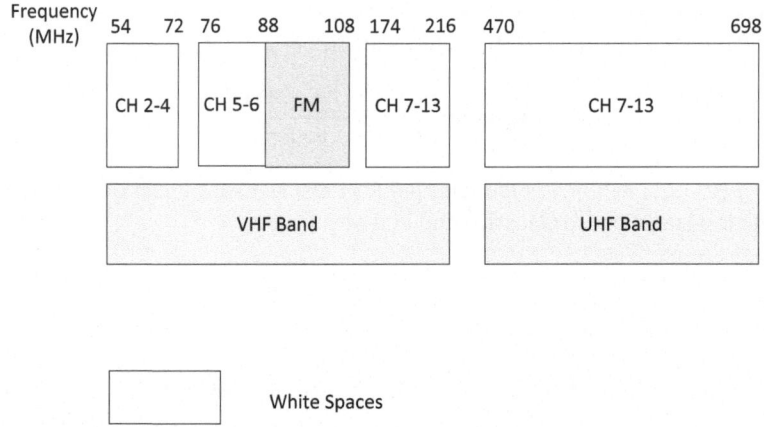

Fig. 1.6 TV white spaces

May 2012 eliminates spectrum sensing as a requisite for cognitive radio devices, and enforces the database query approach. The FCC conditionally designated ten entities as TV white space device database administrators. The conditional database administrators are Spectrum Bridge, Inc., Telcordia Technologies, Inc., Airity, Inc., Comsearch, Frequency Finder, Inc., Google Inc., KB Enterprises LLC and LS Telcom, Key Bridge Global LLC, Microsoft Corporation, NeuStar, Inc. These databases tell TV band devices what channels they can transmit on without causing interference to TV broadcast stations, wireless microphones and other authorized broadcast auxiliary services.

TV white spaces consist of a substantial amount of TV channels are freed up from the transition from analog to digital television broadcasts. The newly freed TV white spaces include the VHF/UHF band from 54 to 698 MHz (channel 2–channel 36, channel 38–channel 51), as shown in Fig. 1.6. In September 2010, the FCC released the final rules [27] to make TV white spaces available for unlicensed access by broadband wireless devices. In FCC's final rule, TV band devices are required to query a geo-location database for vacant TV channels, which is the so called database-driven cognitive radio networks.

A typical database-driven CRN consists of a database service administrator and multiple white space devices (WSDs) [37]. Figure 1.7 illustrates a database-driven CRN in the FCC rules. The database service administrator stores licensed usage information of the TV band in the incumbent database and geo-location information including the coverage area and signal levels of licensed use in the TV/WS database. Besides, the database administrator also holds the following information:

- tolerable interference of the licensed usage,
- the emission characteristics of WSDs,

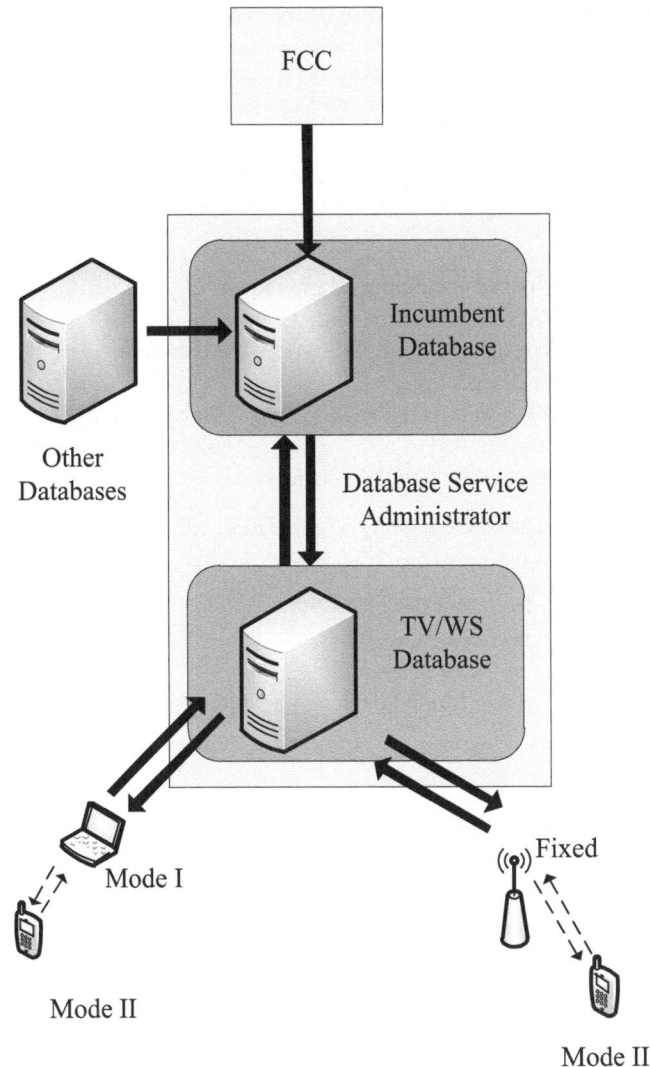

Mode I: Portable device obtains location/channels from fixed device

Mode II: Portable device uses its own geolocation/data base access capability

←‑‑→   Wireless Link

Fig. 1.7 Database-driven TV white space access in the FCC rules

Based on the above information, for a given query containing the location of a WSD, the database can response with

- a list of available channels,
- maximum allowed transmission power.

WSDs are divided into two categories by FCC: Mode I and Mode II [28]. Mode II devices, described as an Enabling STA (station) in IEEE 802.11af specification, are required to

- be capable of identifying its geo-location with granularity of 50 m,
- be able to access the TV/WS database via internet,
- receive available channel/frequency information from the database,
- be able to control its client devices.
- access database periodically (once a day by FCC, every 2 h by Ofcom).
- check its location every 60 s,
- access to the database after moving beyond 100 m.

Mode I devices, described as an Depending STA in IEEE 802.11af specification, follow the following regulations

- does not need to have geo-location capability,
- does not need to have internet access capability,
- does not need to talk to a geo-location database,
- shall be under the control of one Mode I device or a fixed device.

1.2   Privacy Threats in Cognitive Radio Networks

In this section, we will describe what potential privacy issues exist in cognitive radio networks and the significance of protecting users privacy in cognitive radio networks. We will first give some fundamental concepts of location privacy, which has been widely studied in mobile networks. Then, we will extend the notion of privacy to the context of cognitive radio networks. Finally, we will discuss the impacts of privacy information leakage in cognitive radio networks.

1.2.1   Location Privacy

With the proliferation of mobile devices such as smartphones and tablets, location-based services (LBSs) are becoming versatile and improve user's experience by providing personalized information retrieval services based on location-wise information. For example, LBSs can help users find nearest restaurants with their favorite recipes, and keep track of users' physical fitness. Moreover, some of today's popular LBSs also provide location-related information

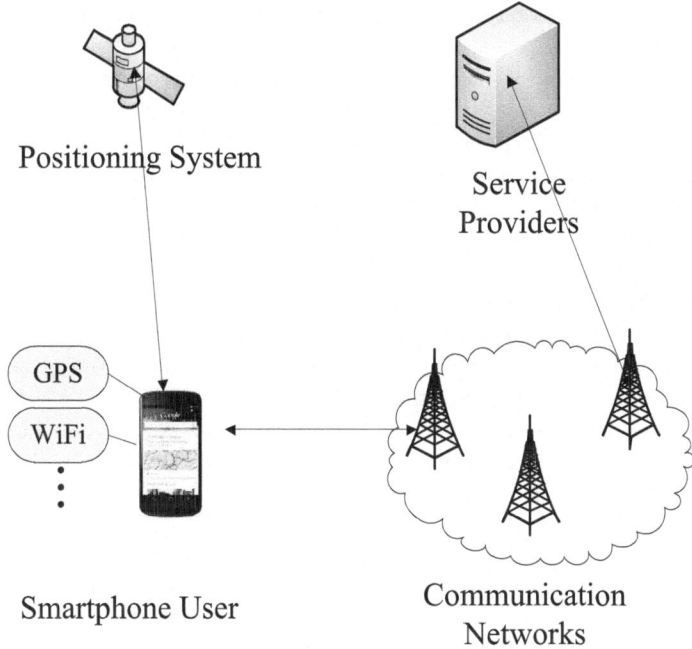

Fig. 1.8 Location-based service system

sharing and social networking features. Foursquare is a good example of such an application that allows users to interact with their nearby friends using mobile devices. Foursquare users can share their real-time location with their friends, explore nearby going-on events, and get reminders based on their locations. Another important application is location-based advertising (LBA) which delivers advertisements to mobile users when they are near a participating retailer or shop.

Typically, as depicted in Fig. 1.8, an LBS system consists of four parts: smartphone user, positioning system, communication networks, and service providers. The users can obtain their location information via positioning system, such as Global Positioning System (GPS), and send location-based queries to LBS service providers via communication networks, e.g., 3G networks or WiFi. The LBS service providers respond the queries based on users' locations.

However, a previous study [51] pointed out that the home location of a driver can be inferred from the GPS data collected on its vehicle even if the data were anonymized or pseudonymized. Another study [68] reported that a variety of personal information such as job title, age, and some personal habits (e.g., coffee drinker, smoker) can be disclosed with indoor location data. Such privacy threats have drawn wide attention from consumers, service providers, researchers, and government organizations. The research project conducted by Beinat [11] shows that 24% of the mobile clients using LBSs have serious privacy concerns on their location traces together with their personal information. A survey [71] conducted by Microsoft also reported that smartphone users concern about the use of their location information. A service provider also has very strong incentives to preserve the location privacy of its users since otherwise the users may choose not to use the LBS. In December 2010, the U.S. Commence Department recommended the inclusion of privacy protection associated with LBSs in electronic privacy laws to improve consumers' privacy protections and to ensure that the Internet remains an engine for innovation and economic growth [93]. Having long been aware of this issue, researchers have proposed many privacy preserving techniques for a variety of scenarios. Typically, LBS service providers are assumed to be honest but curious, meaning that the LBS service providers follow the protocol but try to infer users' private information as much as possible. Such assumption is made based on the consideration that an adversary can be the owner of the LBS server or can compromise the LBS server, in which cases the adversary can access users' information stored at the server as well as the communications between the server and users. Note that under this type of threat model, privacy preservation focuses on how to modify the location information contained in the query. In more complicated cases, the location privacy preservation needs to cope with the adversaries that compromise user's device. Another potential threat comes with an adversary acting as eavesdroppers who passively overhears the communication messages transmission between the LBS server and the users. Existing works on communication security mainly adopt cryptographic techniques to handle such cases. How to secure users' devices and communications against such attacks is security issue which is beyond our discussion of location privacy issue.

Two types of location privacy issues in LBS are considered, namely, query privacy and location privacy, where query privacy is the privacy information contained in the LBS query and location privacy refers to user's private location information. Normally, query privacy issue contains the following two kinds of privacy breach:

- *Identification privacy breach:* The identification of a user is de-anonymized by an adversary.
- *Attribute privacy breach:* An attribute of a user is successfully inferred by an adversary. An attribute describes a property of a user, including user's interest and hobby.

These two types of privacy issues are usually correlated together. If a user has been identified, an adversary can obtain more information from other sources (e.g., the user's historical location related traces) that can be leveraged to locate the user. On the other hand, if a user has been located accurately by the adversary, the adversary can use such knowledge to match the user with its location-information-contained query. Note that it is also possible that only one of these two types of privacy issues is compromised by an adversary.

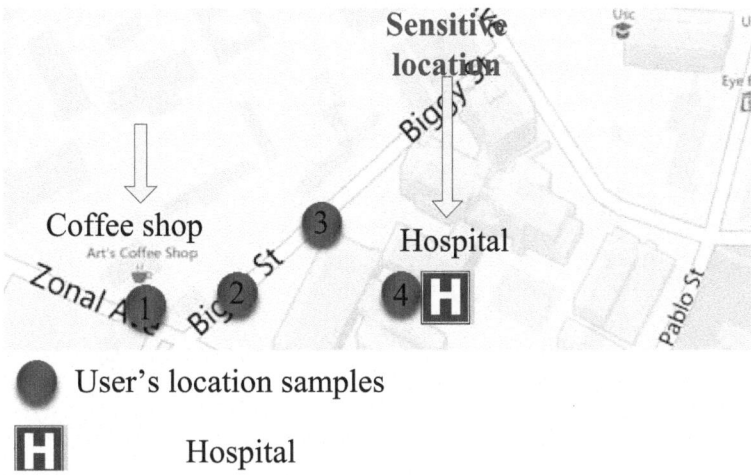

Fig. 1.9 An illustration of location privacy

1.2.2   Location Privacy in Cognitive Radio Networks

In CRNs, location privacy arises with the open nature of wireless communication as well as software defined radio platforms, such as Universal Software Radio Peripherals (USRP) [56] and Sora software radio platform [91]. Nowadays, the growing privacy threats of sharing location information in wireless networks have been widely concerned [106]. The location traces contain a lot personal information about the user. Fine-grained set of location tracking information can be access in the CRNs by correlating sensing reports in collaborative spectrum sensing or directly extracting location information contained in database query.

What's worse, even if a user remove its sensitive visits, the adversaries can still extract them out. For example, as shown in Fig. 1.9, an adversary may learn that a user regularly goes to a hospital by analyzing its location trace pattern (node 1–3), which the user does not want to be disclosed, and sell such information to pharmaceutical advertisers.

In spectrum sensing based CRNs, an SU can sense the signals from multiple PUs, as illustrated in Fig. 1.10. Since the sensing results on the signal strengths are highly correlated to the SU's location (which is the essential observation in received signal strength based localization algorithms), on the other hand, adversaries can link the SU's physical location with its sensing results by adopting certain received signal strength based localization technique.

In database-driven CRNs, an SU's query contains the fine-grained location information with granularity of 50 m, which is quite similar to location based query in mobile networks. Moreover, as stated in [40], the channel usage information over a certain period of time can be leveraged to launch location privacy attack.

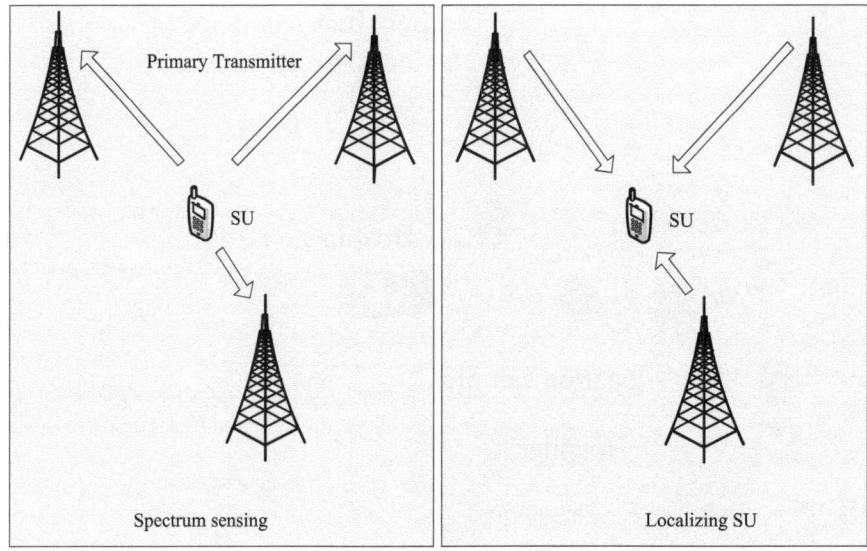

Fig. 1.10 Localizing an SU based on its sensing results

In summary, the SU's sensing results and query contain location information which should be kept private to the SU itself. Unfortunately, in collaborative spectrum sensing, SUs are required to submit their sensing results periodically to the FC for collaboration. In database-driven CRNs, SU needs to update their location information to the database to obtain available channels. In both cases, there are potential privacy risks if there are adversaries in the CRNs.

1.2.3   Significance

Location privacy preserving is very import in our daily lives because it can lead to serious consequences if the personal location information is leaked to adversaries. The fine-grained location data can be used to determine a great deal about an individual's beliefs, preferences, and behavior. Untrusted entities, e.g., compromised FC or databases, may sell such personal information to advertisers without user's permission. For example, a compromised FC may learn that a user regularly goes to a hospital by linking the user's sensing results to its physical location, which the user does not want to be disclosed, and sell such information to pharmaceutical advertisers. Moreover, malicious adversaries with criminal intent could hack the databases or FC with such information to pose a threat to individual security and privacy.

Some real events demonstrate that some publicly known databases can be leveraged to identify a person:

- In 2002, Sweeney showed that about 87% of the patients in a real medical data could be uniquely identified by linking attributes in the data set with a voter registration list.
- In 2006, AOL published a data set without taking enough precautions and thus leaded to some undesirable consequences. The data set published by AOL in 2006 enabled the unique identification of a single 62 year old woman living in Lilburn, Georgia by looking at the search logs by New York Times reporters. The search logs were withdrawn and two responsible employees got fired.
- From the Netflix released moving rating data in 2006, 96% of the subscribers can be uniquely identified by at most eight movie ratings together with rating dates.

The fine-grained location data can be used to determine a great deal about an individual's beliefs, preferences, and behavior. By combining some publicly known databases, the following personal information can be further extracted:

- Economic status can be extracted based on home location. Home location is easy to be deduced from daily location trace. Once the fine-grained home address is known, mortgage balances and tax levies are often easy to derive.
- Individual's beliefs can be deduced by analyzing the user's frequently visit places with religious affiliations.
- Health condition can be determined based on the user's location trace related to clinics and hospitals. The frequency and duration of hospital visits can determine whether the user suffers from a serious illness or even the type of the illness by specialty clinics visits.

Chapter 2
Privacy Preservation Techniques

In this chapter, we will introduce some basic concepts and fundamental knowledge of privacy preservation techniques. Privacy preservation has become a major issue across different applications, from data publishing to location-based services. Although the applications and scenarios are quite different, the essential privacy models and the core techniques are relatively the same. The state-of-the-art privacy techniques can be categorized into four classes: anonymization, perturbation, differential privacy, and cryptographic techniques. In location privacy protection, anonymization boils down into so-called spatial cloaking. Since anonymization and spatial cloaking adopt the same idea, we will discuss them in one section. Cryptography is a very general terminology and covers plenty of specific techniques. In this chapter, we only focus on private information retrieval, which is most related to database query in CRNs. A brief review on the literature of each category will be provided at the end of each section.

2.1 Anonymization and Spatial Cloaking

In this section, we will discuss some fundamental concepts used in anonymization and spatial cloaking. In order to get a better understanding of these privacy preservation techniques, we first illustrate them using a toy example.

2.1.1 Anonymization Operations

we consider six nodes in a network, and each node has the following three attributes

- Identifier (ID) that can uniquely identify a node.
- Longitude.
- Latitude.

W. Wang and Q. Zhang, *Location Privacy Preservation in Cognitive Radio Networks*, SpringerBriefs in Computer Science, DOI 10.1007/978-3-319-01943-7_2, © The Author(s) 2014

Table 2.1 Location data
in the example

	Location data	
Identifier	Longitude	Latitude
1	20	20
2	25	17
3	20	40
4	27	35
5	39	27
6	38	29

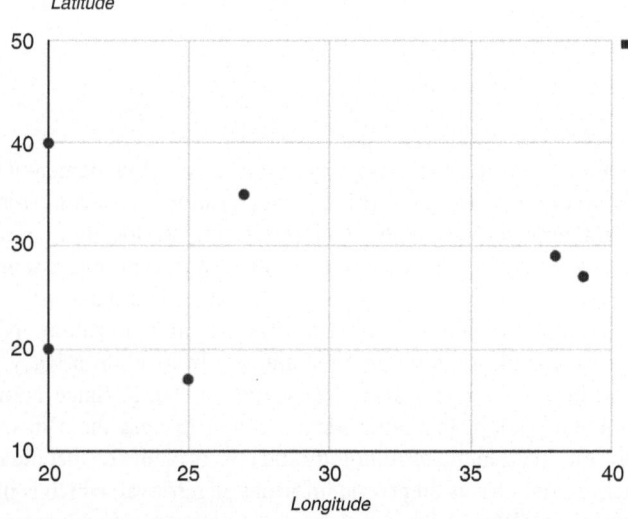

Fig. 2.1 Locations in the example

Longitude and latitude attributes together form a two-dimensional location data that describe the exact two-dimensional geo-location of a node. In a location sensitive scenario, the nodes do not want others to know their exact physical location.

Example 1. Table 2.1 shows an example of location data in this network.

Table 2.1 belongs to relational model of data, where each tuple (e.g., node) attaches a set of attributes (longitude and latitude). As a visual aid, the two-dimensional locations are depicted in Fig. 2.1.

In order to protect node's location privacy, the nodes can *anonymize* or *cloak* their location data before sharing the data with other entities (e.g., the FC or the database). The core idea of anonymization or spatial cloaking consists of the following two steps

- Breaking, which is an operation that divides the nodes into a set of groups and then breaks the exact links between the identifiers and location attributes. Then, the one-to-one mapping between identifiers and locations are weaken by breaking the linkage so that the adversary has less confidence in inferring a node's location. As such,

ID	Longitude	Latitude
1	20	20
2	25	17
3	20	40
4	27	35
5	39	27
6	38	29

Break

ID	Longitude	Latitude
1	20	20
2	25	17
3	20	40

ID	Longitude	Latitude
4	27	35
5	39	27
6	38	29

Grouping

ID	Longitude	Latitude
1	20-25	17-40
2	20-25	17-40
3	20-25	17-40

ID	Longitude	Latitude
4	27-39	27-35
5	27-39	27-35
6	27-39	27-35

Fig. 2.2 Anonymizing the example

- Grouping. Via grouping, all nodes in the same group are indistinguishable in their locations. The most common way to group nodes is generalization, or so-called spatial cloaking in location privacy. Generalization is an operation changing an exact value to a more generalized one, e.g., a numeric value can be generalized to a range value.

Figure 2.2 describes the grouping-and-breaking and generalization operations on Table 2.1. We can see that the nodes are divided into two groups, and the location data in each group is generalized to identical ranges. A two-dimensional range can be viewed as a region or a cloak. The generalization operation to group exact location points into a region is referred to as spatial cloaking, are depicted in Fig. 2.3.

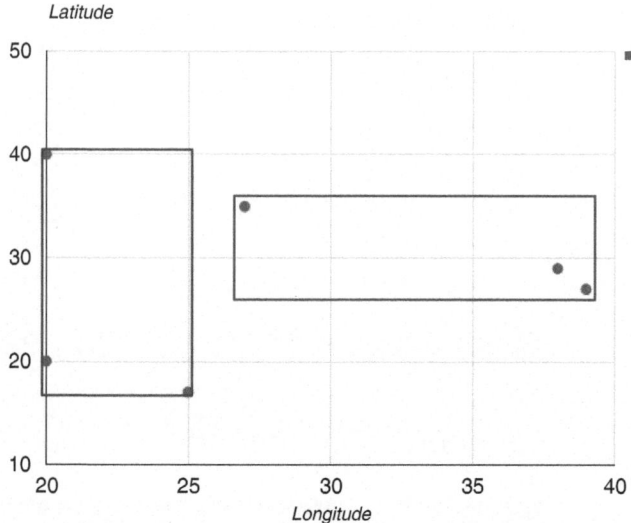

Fig. 2.3 Spatial cloaking over the example

As mentioned in introduction, location privacy consists of two categories: query privacy and location privacy. Here we use an example in LBS to show the difference of using k-anonymity to protect query privacy and location privacy.

Example 2. Similar to the above example, k-anonymity is applied to the location information contained in the query so that the exact locations of k users are generalized to the same region. In this way, the identity of each user is protected since all users are indistinguishable from each other by looking at the queries. Thus, query privacy is well protected. However, when the k users locate in a very small region, an adversary can locate the users in a restricted area without knowing the identity of each query. An extreme case is that all users co-locate together, i.e., all users stand in the same location spot. In such situation, the location privacy of the users are compromised.

In the above example we can see that even if k-anonymity is applied, the location privacy can still be compromised. To address this issue, location entropy, inspired by Shannon's entropy in information theory, is proposed to measure privacy level. Location entropy describes adversaries' uncertainty about a user's location. Normally, a set of points of interest (POI) is specified, and is usually assumed that the adversaries have equal inference beliefs of a user visiting a POI. Note that with equal beliefs, the location entropy is the largest, meaning that the adversaries' uncertainty is the highest before observing the location queries. The posterior belief entropy is used to measure the uncertainty about a user's location after observing the location queries.

A major limitation of location entropy is that the choice of POI affect the result of entropy. Another issue is that location entropy cannot tell how accurately an adversary can infer a user's location. To cope with these issues, expected distance error is used as a improved version location entropy, where the location error of an inferred location is taken into consideration. The location error refers to the distance between the inferred location and the true location of a user. Combining inference beliefs, expected location error can be computed. As such, different POIs have different location error and thus weight different in the measurement.

The information loss in the process of anonymization is inevitable, which is a tradeoff for protecting privacy. To quantify the information loss caused by anonymization operations, several information metrics are developed. In the following part of this subsection, we will introduce several information loss metrics.

A simple measure on information loss is minimal distortion (MD), where a unit penalty is counted when each element in the data table is generalized or modified. For example, in Fig. 2.2, a total of 12 elements (i.e., 6 individuals, and each individual has 2 elements generalized) have been generalized, and thus, 12 units of distortion will be counted. MD is usually suitable for operations on categorical element, while for operations on numerical elements, information loss and discernibility metric are commonly used.

Information loss, or $ILoss$, first proposed in [108], is a metric capturing the information loss of generalizing a specific value to a general range

$$ILoss(R(v_A)) = \frac{R(v_A)}{D(A)}, \tag{2.1}$$

where v_A is an exact value of attribute A, $R(v_A)$ the generalized range of the value v_A, $D(A)$ the domain of the attribute A. $ILoss(R(v_A))$ measures the information loss caused by generalizing v_A to $R(v_A)$, i.e., the fraction of domain values generalized by $R(v_A)$. For instance, if the domain of a location data is [1, 60], generalizing the location 5 to a range [1, 10] has information loss of $(10 - 1)/60$. The total information loss of an individual u can be computed by

$$ILoss(u) = \sum_{A \in \mathscr{A}} (\omega_A ILoss(R(v_A))), \tag{2.2}$$

where ω_A specifies the importance of the attribute A, $\mathscr{A}$ the set of all attributes. The overall loss can be measured by

$$ILoss(\mathscr{U}) = \sum_{u \in \mathscr{U}} ILoss(u), \tag{2.3}$$

where $\mathscr{U}$ is the set of all individuals.

The normally used metric for k-anonymity is discernibility metric (DM) [52]. The aim of DM is to quantify the loss for each individual being anonymized to be indistinguishable from other individuals with respect to QIs. Let $|QI_i|$ denote the size of a QI group i. Then, the penalty of generalizing this group is $|QI_i|^2$. The overall penalty can be expressed as

$$ILoss(\mathscr{U}) = \sum_i |QI_i|^2. \tag{2.4}$$

2.1.2   Anonymization Privacy Models

In the previous subsection we discussed the fundamental operations in anonymization technique. In this subsection, we will introduce some anonymization privacy models to quantitatively analyze location privacy protection.

As the first and the most fundamental anonymization privacy model, k-anonymity [89] has been proposed to protect individual privacy in data publishing.

Definition 1 (k-Anonymity). In each group partitioned by a certain anonymization algorithm is said to satisfy k-anonymity if the size of the group equals or is larger than k. A table is said to satisfy k-anonymity if each group in the table satisfies k-anonymity.

The goal of k-anonymity is to ensure that each individual is indistinguishable from at least $k - 1$ other individuals in the table. In a location sensitive network, k-anonymity is satisfied if each node's shared location in the network is indistinguishable from at least $k - 1$ other nodes' locations. In Fig. 2.2, we can see that the table is 3-anonymity: locations of node 1, 2, 3 are generalized to the same region and thus cannot distinguish from each other, and so do locations of node 4, 5, 6.

Based on the notion of k-anonymity, many other anonymization models have been proposed. Here we introduce several popular anonymization models, including ℓ-diversity [66], (α, k)-anonymity [107].

The essential observation of ℓ-diversity is that the attributes of an individual consists of two parts: quasi-identifiers (QIs) and sensitive attributes (SAs). QIs refer to the attributes that can semi-determine the individual, and usually correspond to personal information such as gender, nationality, and age. QIs can be used to identify an individual. A previous study [89] shows that 87 % individuals in a medical data set could be uniquely identified by gender, date of birth and zip code. These QIs are accessible from some publicly known data set; examples include a voting registration table. According to Sweeney [89], most municipalities sell the identifiers of individuals along with basic demographics, including local census data, voter lists, city directories, and information from motor vehicle agencies, tax assessors, and real estate agencies; the study [89] also points out that a city's voter list in two diskettes was purchased for only 20 dollars, and was used to identify medical records. QIs are usually considered to be publicly known and are treated as background knowledge of the adversaries. In location privacy, the longitude and latitude attributes can be considered as QIs since they can identify a user by its physical location.

SAs are the attributes that are private to the individuals, such as home address in a location application or disease attributes in a medical data set. Given QIs as background knowledge, k-anonymity may not be able to protect individual's privacy. Consider Table 2.1 again with one more attribute "AtHome" added. 'AtHome" is sensitive to all users, as described in Table 2.2. After applying anonymization algorithm on Table 2.2, we can see the k-anonymous table with sensitive attribute "AtHome" as depicted in Fig. 2.4. We can see that although the

Table 2.2 Location data
with "AtHome" attribute in
the example

Identifier	Location data (QI)		SA
	Longitude	Latitude	AtHome
1	20	20	Y
2	25	17	Y
3	20	40	Y
4	27	35	N
5	39	27	N
6	38	29	N

ID	Longitude	Latitude	AtHome
1	20-25	17-40	Y
2	20-25	17-40	Y
3	20-25	17-40	Y

ID	Longitude	Latitude	AtHome
4	27-39	27-35	N
5	27-39	27-35	N
6	27-39	27-35	N

Fig. 2.4 k-anonymous table with sensitive attribute

exact location of users are anonymized, we can tell whether a user is at home or not
since in each group the values of the "AtHome" attribute are identical. To tackle this
case, ℓ-diversity is proposed [66].

Definition 2 (ℓ-Diversity). In each group partitioned by a certain anonymization
algorithm is said to satisfy ℓ-diversity if the probability that any individual in this
group is linked to a sensitive value is at most $1/\ell$. A table is said to satisfy ℓ-diversity
if each group in the table satisfies ℓ-diversity.

By definition, we can see that the example in Fig. 2.4 only satisfies 1-diversity.
To improve the diversity of sensitive values, we anonymize the table with a different
partition strategy as illustrated in Fig. 2.5. In Fig. 2.5, it can be seen that each
sensitive value in a group has frequency no more than 2 and the group size is fixed at
3, thus, the result satisfies 2/3-diversity as well as 3-anonymity. Figure 2.6 depicts
the spatial cloaking results.

An important property shared by the aforementioned two anonymization models
is *monotonicity* property, which is commonly used as an essential observation
to design anonymization algorithms. The definition of monotonicity is given as
follows.

Definition 3 (Monotonicity). A privacy model $\mathcal{M}$ is said to satisfy the monotonicity property, if for any two groups G_1 and G_2 that satisfy $\mathcal{M}$, the merged group
$G_1 \cup G_2$ satisfies $\mathcal{M}$.

ID
1
2
5

Longitude	Latitude	AtHome
20-39	17-27	Y
20-39	17-27	Y
20-39	17-27	N

ID
3
4
6

Longitude	Latitude	AtHome
20-39	29-40	Y
20-39	29-40	N
20-39	29-40	N

Fig. 2.5 Another k-anonymous table ℓ-diversity considered

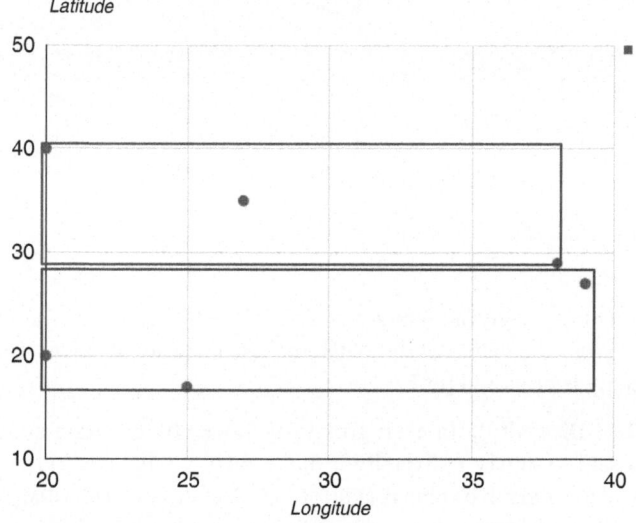

Fig. 2.6 Spatial cloaking with ℓ-diversity considered

Both k-anonymity and ℓ-diversity satisfy the monotonicity property. Taking Fig. 2.5 as an illustration. In Fig. 2.5, each of the two groups satisfies 2/3-diversity and 3-anonymity. If we merge the two groups into one single group containing all six individuals, the resulting group still satisfies 2/3-diversity and 3-anonymity.

Besides the above-mentioned anonymization models, there are many alternative models, most of which can be considered as the enhanced or variant versions of k-anonymity or ℓ-diversity.

Jointly considering the requirements of k-anonymity and ℓ-diversity, (α, k)-anonymity [107] is proposed. In (α, k)-anonymity model, α is a real number within the interval $[0, 1]$ and k is a positive integer. A table is said to satisfy (α, k)-anonymity model if the size of each group in the table is at least k and the frequency of each sensitive value in a group is at most αk. The interpretation of (α, k)-anonymity model can be the combination of ℓ-diversity and k-anonymity where α is set to $1/\ell$.

A variant version of ℓ-diversity is entropy ℓ-diversity, where instead of using frequency to bound sensitive values, entropy is adopted.

$$- \sum_{s \in \mathscr{S}} \Pr(s) \log \Pr(s) \geq \log(\ell), \tag{2.5}$$

where s is a sensitive value, $\mathscr{S}$ the set of all possible values of a sensitive attribute, $\Pr(s)$ the fraction of individuals in a group with sensitive value s. The left hand side of the equation is called the entropy of the sensitive attribute. The purpose of entropy ℓ-diversity is to generate a result with more evenly distributed sensitive values in each group.

To cope with high-dimensional data table, Mohammed et al. [76] reported that in real-life attacks, the adversaries could hardly acquire all QIs. Based on this observation, Mohammed et al. proposed (LKC)-privacy, where the prior knowledge of adversaries is assumed to be limited to at most L QI attributes, K and C are the bounds similar to k-anonymity and ℓ-diversity.

The notion of t-Closeness is proposed [63] to maintain the overall distribution of a sensitive attribute. Consider a case where 95% of individuals are in a coffee shop area while only 5% are scattered elsewhere. Suppose a group with 50% of individuals in a coffee shop and 50% elsewhere, and therefore satisfies 2-diversity. However, this group leaks extra information on location spots since any individual in this group could be inferred as being elsewhere with much higher confidence compared with overall probability of being elsewhere. To prevent such information leakage, t-Closeness [63] requires the distribution of a sensitive attribute in any group to be close to the distribution of the attribute in the overall data set. Earth Mover Distance (EMD) function is leveraged to measure the closeness between two distributions, and the closeness is bounded to be within t.

2.2 Random Perturbation

Random perturbation [4, 105] is a popular method for eliciting information from individuals without compromising privacy, which has been adopted in many privacy preservation applications such as advertisement targeting [57], data mining [33, 109], collaborative sensing [65], collaborative spectrum sensing [64] due to its simplicity, efficiency, and statistical preservation. The basic idea of random perturbation is to replace the original data values with some synthetic data values so that the statistical information remains relatively the same while the original values never get disclosed. Due to the randomization, an individual perturbed value can be quite different from its original value. As such, the true values are kept private and cannot be inferred by the adversaries by linking private attributes to a certain individual.

There are many different random perturbation techniques. First, we illustrate the core idea of random perturbation by introducing a specific random perturbation which is referred to as uniform perturbation. The principle of uniform perturbation is as follows: for each sensitive value v, we toss a coin with head probability of p and tail probability of $1 - p$. If the tossing result is head, v remains unchanged; otherwise v is replaced with a random value sampled from its domain. For example, $v = 3$, $p = 0.6$, the domain of v is integers within $[1, 10]$, then the mapping result of v, denoted as $\hat{v}$, can be written as

$$\hat{v} = \begin{cases} v & \text{with probability } 0.6 \\ X & \text{with probability } 0.4 \end{cases} \tag{2.6}$$

where X is a random variable follows a discrete uniform distribution that each value in the domain is chosen with equal probability. Then, the probability that v is retained is $0.6 + 0.4 \times 1/10 = 0.64$. The probability mass function (pmf) of $\hat{v}$ is given by

$$pmf(\hat{v}) = \begin{cases} 0.64, & \text{where } \hat{v} = v \\ 0.04, & \text{otherwise} \end{cases} \tag{2.7}$$

Besides uniform perturbation, there are many other perturbation models with different perturbation distribution. For example, (p, γ)-perturbation is proposed to conceal user's interest in a certain advertisement. The user's interest can be interpreted as a binary variable B where $B = 1$ indicates that the user is interested in the advertisement while $B = 0$ indicates that the user is not interested in the advertisement. (p, γ)-perturbation can be expressed by the following formula

$$\hat{B} = \begin{cases} B, & \text{with probability } p, \\ 1, & \text{with probability } (1 - p)\gamma, \\ 0, & \text{with probability } (1 - p)(1 - \gamma) \end{cases} \tag{2.8}$$

Similar to uniform perturbation, the implementation of (p, γ)-perturbation can be considered as tossing two biased coins with head probability of p and γ respectively. Each user can pick p and γ by one of the following two rules:

- *Fixed rule.* Each user can use fixed p and γ to perturb the sensitive values. And the fixed p and γ is known by the adversary.
- *Randomized rule.* Each user picks the values for p and γ from some certain known distributions. The distributions for p and γ are independent with each other and can be different.

The above mentioned perturbation techniques utilize certain kinds of distribution over the domain of sensitive value to perturb the original value with a pre-defined probabilities. It is easy to see that these techniques are suitable for attributes with discrete domain to randomly generate values from probability mass functions. As for numerical values, such as longitude and latitude attributes, additive noise is normally used to perturb the original values.

The general idea of additive noise is to add a random noise δ to the original value v and replace v with the noisy result $\hat{v} = v + \delta$. Typically, the random noise δ is drawn from some distributions independently so that the overall statistical properties, such as means and correlations, are retained. This class of methods gains much popularity in the communities like data mining and participatory sensing, where people care about the overall statistical information rather than individual's record.

A major drawback of additive noise is that it usually needs a large amount of noise to preserve privacy, and the random noise overwhelms the original features contained in the true data. A smart way to cope with this challenge is to project the original data to another space while some features and relationships in the original space are retained. Noise or randomness can be injected into the projection process to preserve privacy so that the original data cannot be recovered by an invert process.

A loss transformation on data can be a projection that converts the original data to a lower dimensional space since in general this type of projection is not reservable. The basic observation is that a high dimensional data point can be uniqueness projected to a lower dimensional data point while it is usually not possible to recover the high dimensional data point only using the lower dimensional data point, since there is information loss in the process of projection. Thus, this type of projection is adopted in [65, 116], and usually follows three steps:

- Choose a dimensionality d, which is lower than the original dimensionality m;
- Construct a projection matrix $\mathbf{A}$ of size $d \times m$;
- Project each data point $\mathbf{x} \in \mathbb{R}^m$ into $\mathbf{z} \in \mathbb{R}^d$ via the projection $\mathbf{z} = \mathbf{A}^\top \mathbf{x}$.

Even if the projection matrix $\mathbf{A}$ is known by the adversary, it is usually not possible to recover $\mathbf{x}$ from $\mathbf{z}$. Thus, the privacy information in $\mathbf{x}$ is not visible in $\mathbf{z}$. The key challenge to enable such technique is the construction of the projection matrix $\mathbf{A}$ so that $\mathbf{z}$ still conveys the same useful information as $\mathbf{x}$. As such, additional ingredients are put onto the projection matrix to define optimality or usefulness of the results. In [85, 116], the relative distance between data points are kept, i.e., among any three data points $\mathbf{x}_i, \mathbf{x}_j, \mathbf{x}_k$, the relationship that $\mathbf{x}_j$ is more similar to $\mathbf{x}_i$ than $\mathbf{x}_k$ is retained after projection. This type of constraint can be written as

$$\|\|\mathbf{A}^\top(\mathbf{x}_i - \mathbf{x}_j)\|\|_2^2 \le \|\|\mathbf{A}^\top(\mathbf{x}_i - \mathbf{x}_k)\|\|_2^2,$$

$$\forall i, j, k \in \left\{ i, j, k : \|\|(\mathbf{x}_i - \mathbf{x}_j)\|\|_2^2 \le \|\|(\mathbf{x}_i - \mathbf{x}_k)\|\|_2^2 \right\} \tag{2.9}$$

However, some important relationships between the feature vectors, such as Euclidean distances and inner products, are lost in the process of dimensionality-reducing transformation. For example, the inner product $\mathbf{x}_i^\top \mathbf{x}_j$ is not identical to $\mathbf{x}_i^\top \mathbf{A} \mathbf{A}^\top \mathbf{x}_j$ unless $\mathbf{A}$ is an orthogonal matrix, which, however, necessarily preserves dimensionality. Therefore, distortion introduced by dimensionality reduction can lead to the performance degradation in using the projected data for certain tasks.

Instead of employing dimensionality-reducing projection, random noises are injected into the projection to preserve privacy. Pickle [65] is a most recent work that applies such technique to participatory sensing. The target of Pickle uses

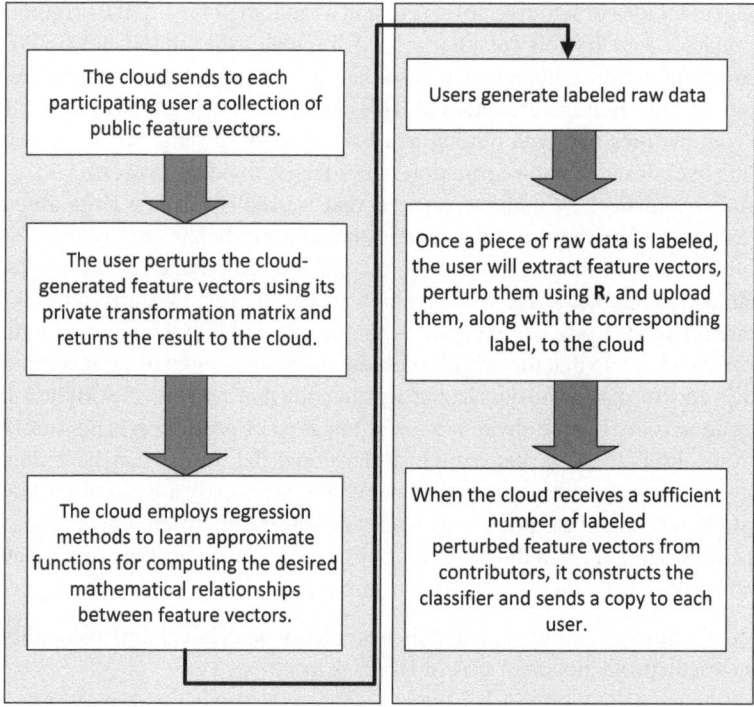

Fig. 2.7 Work flow of Pickle

participatory sensing to build a classifier while keeping each contributor's data private. Pickle first learns a statistical model to compensate the data distortion in the projection. And then Pickle reconstructs the original inner products and distance from the perturbed data, based on which Pickle constructs the final classifier. The work flow of Pickle is illustrated in Fig. 2.7.

The key intuition in Pickle is using regression (the first three blocks in Fig. 2.7) to learn the most important covariance structures in the underlying training data without being able to regenerate the original feature vectors.

In the first step, the cloud randomly generates public feature matrix $\mathbf{Z}$ and sends $\mathbf{Z}$ to each user. Then, each user perturbs $\mathbf{Z}$ in exactly the same way as the user would perturb the actual training feature vectors. To do this, each user keeps a private random matrix $\mathbf{R}_u$ to perturb $\mathbf{Z}$ into $\mathbf{R}_u^{\top}\mathbf{Z}$. However, this approach has the following vulnerability. If the private $\mathbf{R}_u$ can be recovered as long as $\mathbf{Z}$ is invertible, i.e., the cloud can recover $\mathbf{R}_u$ by computing $\mathbf{R}_u^{\top}\mathbf{Z}\mathbf{Z}^{-1}$ (note that $\mathbf{Z}$ is known by the cloud). This would lead to privacy disclosure. To address this issue, each user adds an additive random noise matrix $\boldsymbol{\delta}_u$ to $\mathbf{Z}$. Note that although $\mathbf{Z}$ is publicly known, $\boldsymbol{\delta}_u$ is kept private to each user. Then, the cloud would recover the original relationships from the perturbed feature vectors. After learning the original relationships, it is ready to construct pattern classifiers using training samples contributed by users.

2.2.1 Privacy Measure

Since random perturbation does not focus on the exact individual or attribute that the adversary can successfully attack by linking its sensitive attribute with the true value, but concentrates on how to camouflage the true values by replacing them with some random variables. The random variables have impacts on the adversaries beliefs on the true values. The intuitive measure on privacy level is to quantify the information that can be disclosed by how much it would change the adversaries' beliefs on the true values to observe the perturbed values. Generally, this type of privacy measures try to ensure privacy protection by limiting the difference between the prior and posterior beliefs on the true values. In the following part, we will briefly describe some privacy measures for random perturbation.

Uniform perturbation can provide privacy guarantee measured by the difference in the adversary's prior and posterior beliefs. $\rho_1 - \rho_2$ privacy [36] and $\delta - growth$ [94] are two privacy models using such privacy measure. Let $\Pr[v]$ and $\Pr[v|\hat{v}]$ denote the prior and posterior beliefs on the sensitive value v before/after observing $\hat{v}$. $\rho_1 - \rho_2$ privacy bounds the prior and posterior beliefs by

$$\Pr[v] < \rho_1 \Rightarrow \Pr[v|\hat{v}] < \rho_2,$$
$$\text{and} \quad \Pr[v] > \rho_2 \Rightarrow \Pr[v|\hat{v}] > \rho_1, \tag{2.10}$$

where $0 < \rho_1 < \rho_2 \leq 1$.

Similarly, $\delta - growth$ bounds the prior and posterior beliefs by

$$\Pr[v|\hat{v}] - \Pr[v] < \delta, \tag{2.11}$$

where $0 < \delta < 1$ bounds the growth in the belief on sensitive value v by observing $\hat{v}$.

As an adaptation of $\rho_1 - \rho_2$ privacy, (d, γ)-privacy is proposed in [84] to bound difference in prior and posterior beliefs, and provides a provable guarantee on privacy as well as utility. Let $\Pr[v]$ and $\Pr[v|\hat{v}]$ denote the prior and posterior beliefs on the sensitive value v before/after observing $\hat{v}$. The formal definition of (d, γ)-privacy is given as follows.

Definition 4 ((d, γ)-Privacy). Let $\Pr[v]$ and $\Pr[v|\hat{v}]$ denote the prior and posterior beliefs on the sensitive value v before/after observing $\hat{v}$. A random algorithm is (d, γ)-private if the following conditions hold for all d-independent adversaries

$$\frac{d}{\gamma} \leq \frac{\Pr[v|\hat{v}]}{\Pr[v] \leq d}, \tag{2.12}$$
$$\text{and } \Pr[v|\hat{v}] \leq \gamma.$$

Where adversaries are d-independent if the prior belief satisfies the conditions $\Pr[v] \leq d$ or $\Pr[v] = 1$ (d, γ)-privacy achieves a reasonable trade-off between privacy and utility when the prior belief is small.

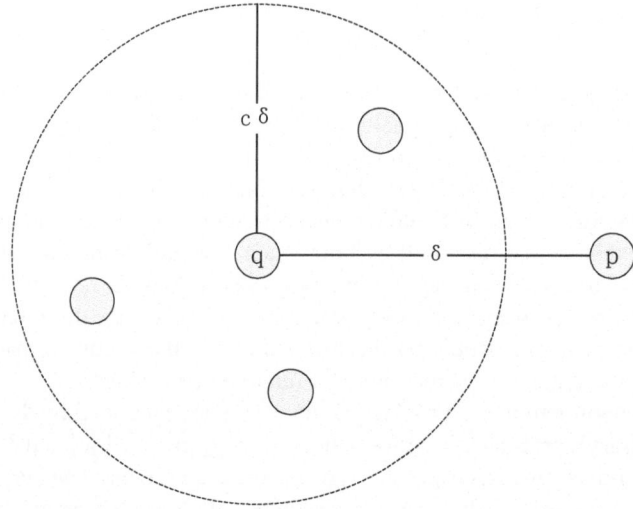

Fig. 2.8 (c, t)-isolation

It is reported in [14] that privacy can be interpreted as the adversary's power of isolating an individual should not be enhanced by observing the released results. Based on this intuition, (c, t)-isolation is proposed. Denote the original data point as p, the corresponding data point inferred by adversary as q, and δ as the distance between p, q. As illustrated in Fig. 2.8, it is said that q (c, t)-isolates p if there are fewer than t data points falling into a ball of radius δ centered at q. This model uses distance to measure privacy level, which is suitable for projection based perturbation.

2.3 Differential Privacy

Differential privacy is the most popular privacy model used in database systems. In this section, we first introduce some basic concepts in the differential privacy model. Then, we interpret differential privacy in terms of priori and posterior beliefs. Finally, we review the state-of-the-arts that are related to location data.

2.3.1 Differential Privacy Model

We first introduce some basic concepts of differential privacy. The intuition of differential privacy is that the removal or addition of a single record does not significantly affect the outcome of any analysis. The following is the formal definition of ϵ-differential privacy in the *non-interactive setting* [12], where ϵ specifies the degree of privacy ensured.

Definition 5 (ϵ-Differential Privacy). A mechanism $\mathcal{M}$ provides ϵ-differential privacy for an SU u if for any possible sets of sensing reports $\mathbf{R} = [\mathbf{r}_1, \ldots, \mathbf{r}_u, \ldots, \mathbf{r}_U]$ and $\mathbf{R}' = [\mathbf{r}_1, \ldots, \mathbf{r}'_u, \ldots, \mathbf{r}_U]$ differing only on u's sensing data,

$$\left| \ln \frac{\Pr[\mathcal{M}(\mathbf{R}) = \mathbf{O}]}{\Pr[\mathcal{M}(\mathbf{R}') = \mathbf{O}]} \right| \leq \epsilon, \tag{2.13}$$

for all $\mathbf{O} \in Range(\mathcal{M})$, where $Range(\mathcal{M})$ is the set of possible outputs of $\mathcal{M}$.

The parameter $\epsilon > 0$ specifies the level of privacy. Specifically, lower value of ϵ ensures stronger privacy. Normally, ϵ is set to be small enough (e.g., 0.1) to make sure that $\Pr[\mathcal{M}(\mathbf{R}) = O]$ and $\Pr[\mathcal{M}(\mathbf{R}') = O]$ are roughly the same, meaning that the output O is insensitive to the change of any single individual's data.

From the viewpoint of an adversary, (2.13) can be rewritten as

$$\left| \ln \frac{\Pr[\mathbf{O}|\mathbf{R}, \mathcal{M}]}{\Pr[\mathbf{O}|\mathbf{R}', \mathcal{M}]} \right| \leq \epsilon. \tag{2.14}$$

We denote $\hat{\mathbf{R}} = [\mathbf{r}_1, \ldots, \mathbf{r}_{u-1}, \mathbf{r}_{u+1} \ldots, \mathbf{r}_U]$, and assume that each SU's sensing data is independent of each other [18,74]. Applying *Bayesian rule* on the LHS of (2.14), we have

$$\left| \ln \frac{\Pr[\mathbf{O}|\mathbf{R}, \mathcal{M}]}{\Pr[\mathbf{O}|\mathbf{R}', \mathcal{M}]} \right| = \left| \ln \frac{\Pr[\mathbf{r}_u|\mathbf{O}, \hat{\mathbf{R}}, \mathcal{M}] \Pr[\mathbf{r}'_u|\hat{\mathbf{R}}, \mathcal{M}]}{\Pr[\mathbf{r}'_u|\mathbf{O}, \hat{\mathbf{R}}, \mathcal{M}] \Pr[\mathbf{r}_u|\hat{\mathbf{R}}, \mathcal{M}]} \right| = \left| \ln \frac{\Pr[\mathbf{r}_u|\mathbf{O}] \Pr[\mathbf{r}'_u]}{\Pr[\mathbf{r}'_u|\mathbf{O}] \Pr[\mathbf{r}_u]} \right|. \tag{2.15}$$

Combining (2.14) and (2.15), we derive

$$e^{-\epsilon} \cdot \frac{\Pr[\mathbf{r}_u]}{\Pr[\mathbf{r}'_u]} \leq \frac{\Pr[\mathbf{r}_u|\mathbf{O}]}{\Pr[\mathbf{r}'_u|\mathbf{O}]} \leq e^{\epsilon} \cdot \frac{\Pr[\mathbf{r}_u]}{\Pr[\mathbf{r}'_u]}. \tag{2.16}$$

$e^{-\epsilon}$ and e^{ϵ} approach 1 as ϵ decreases, which implies that adversaries obtain roughly no extra information about SU's sensing data by observing $\mathbf{O}$, given the condition that ϵ is small enough.

The standard mechanism to achieve differential privacy utilizes the *sensitivity of a mapping*, which is defined as follows:

Definition 6 (Sensitivity of a Mapping). For any mapping $f : D \rightarrow \mathbb{R}^d$, the sensitivity of f is

$$\Delta f \triangleq \max_{\mathbf{D}, \mathbf{D}'} \| f(\mathbf{D}) - f(\mathbf{D}') \|_1, \tag{2.17}$$

for all input matrices $\mathbf{D}, \mathbf{D}'$ differing at most one user's record.

To ensure that an output $f(\mathbf{D})$ is ϵ-differential private, one standard mechanism [34] is to add random noise to $f(\mathbf{D})$, such that the noise follows a zero-mean Laplace distribution with noise scale of $\frac{\Delta f}{\epsilon}$, denoted as $\mathrm{Lap}(\frac{\Delta f}{\epsilon})$.

One principle mechanism to achieve differential privacy is exponential mechanism [69], which is suitable for algorithms whose outputs are not real numbers or make no sense after adding noise. The exponential mechanism selects an output from the output domain, $o \in \mathcal{O}$, that is close to the optimum with respect to a utility function while preserving differential privacy. The exponential mechanism assigns a real valued utility score to each output $o \in \mathcal{O}$, where outputs of higher scores are assigned with exponentially greater probabilities. Let the sensitivity of the utility function be $\triangle u = \max_{T,T',o} |u(T,o) - u(T',o)|$. The probability associated with each output is proportional to $\exp\left(\frac{\epsilon u(T,o)}{2 \triangle u}\right)$.

Theorem 1 ([69]). *Given a utility function $u : (T \times \mathcal{O}) \rightarrow \mathbb{R}$, a mechanism $\mathscr{A}$ that selects an output o with probability proportional to $\exp\left(\frac{\epsilon u(T,o)}{2 \triangle u}\right)$ satisfies ϵ-differential privacy.*

Another generally used mechanism that fits into differential privacy is Laplace mechanism. Laplace mechanism is suitable for the case where the attribute values are real numbers. The standard way to use Laplace mechanism to achieve differential privacy is to add Laplace noise to the original output of a function. Formally, taking the data set $\mathbf{D}$, a function $f : D \rightarrow \mathbb{R}^d$, and the differential privacy parameter ϵ as inputs, the additive Laplace noise follows the p.d.f.

$$p.d.f.(x|\lambda) = \frac{1}{2\lambda} e^{|x|/\lambda}, \tag{2.18}$$

where x is the magnitude of the additive noise, λ is a parameter determined by the sensitivity of f as well as differential privacy parameter ϵ.

Theorem 2 ([34]). *Given a function $f : D \rightarrow \mathbb{R}^d$ over an arbitrary input data set D, the mechanism $\mathscr{A}$ that modifies the output of f with the following formula*

$$\mathscr{A}(D) = f(D) + Lap(\Delta f/\epsilon) \tag{2.19}$$

satisfies ϵ-differential privacy, where $Lap(\Delta f/\epsilon)$ draws from a laplace distribution with $p.d.f.(x|(\Delta f/\epsilon))$.

To analyze the privacy level in a sequence of operations, the composability of differential privacy is introduced [70], which ensures privacy guarantees for a sequence of differentially private computations. For a sequence of computations that each provides differential privacy in isolation, the privacy parameter values add up, which is referred to as *sequential composition*. The properties of sequential composition is stated as follows.

Theorem 3 ([70]). *Let each computation A_i provides ϵ_i-differential privacy. The sequence of A_i provides $\sum_i \epsilon_i$-differential privacy.*

In a special case that the computations operate on disjoint subsets of the data set, the overall privacy guarantee depends only on the worst of the guarantees of each computation, not the accumulation. This is known as *parallel composition*.

Theorem 4 ([70]). *Let each computation A_i provides ϵ_i-differential privacy for each D_i, where D_i are disjoint subsets of the original data set. Then, the sequence of A_i provides $(\max(\epsilon_i))$-differential privacy.*

Recently differential privacy has gained considerable attention as a substitute for anonymization approaches. Numerous approaches are proposed for enforcing ϵ-differential privacy in releasing different types of data. Several data types are related with the data in location privacy. For example, location data and sensing data from a user can be considered as a tuple in histogram data or contingency data table; a location-based query can be considered as a item set in set-valued data. In the following part of this section, we will review several differential privacy approaches on different data types.

2.3.2 Applying Differentially-Private to Set-Valued Data

Publishing different types of data is studied, such as set-valued data [17], decision trees [38], frequent items in transaction data [95], search logs [45, 58], and aggregated results on time series data [83]. These types of data are generally referred to as set-valued data in which each individual is associated with a set of items drawn from a discrete domain. The following table is an example of a set-valued data, where I_i is an item, and the discrete domain of items is $D_I = \{I_1, I_2, I_3, I_4\}$. A taxonomy tree can be used to present all possible generalizations of a item set. A context-free taxonomy tree, which is defined in [17], is a taxonomy tree where the internal nodes the tree are sets of multiple leaves. Note that "context-free" means that the internal nodes in the tree do not have any physical meanings. Figure 2.9 presents a context-free taxonomy tree for items with domain $D_I = \{I_1, I_2, I_3, I_4\}$. According to the taxonomy tree, an item can be generalized to one of its ancestors in the taxonomy tree. We can see that in this example, an item can be generalized to an internal node that contains the item, e.g., I_2 can be generalized to $I_{\{1,2\}}$ but not $I_{\{3,4\}}$.

Intuitively, to release a set-valued data set with differential privacy, a naive method can be directly adding Laplace noise to each possible query on the set-valued data set: first, generating all possible item sets from the item domain, then counting the appearance of each item set in the data set, and adding independent random Laplace noise to each count. However, this simple method has two main drawbacks that make it infeasible to implement: (1) The number of all possible item sets grow exponentially with the number of possible items, so it cannot scale to large data set. (2) the additive noise scale to a new item set accumulates exponentially, which makes the results overwhelmed by noise in a large data set.

Fig. 2.9 A context-free
taxonomy tree

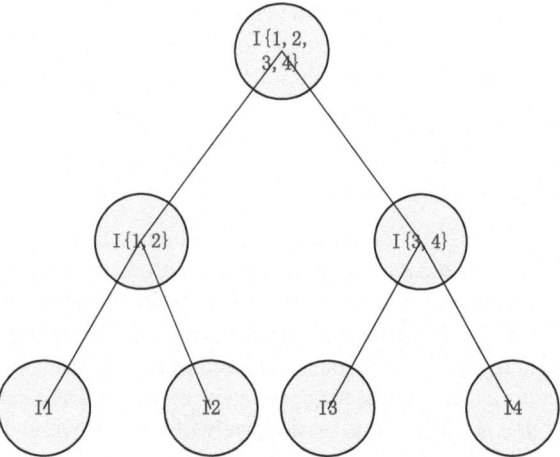

Table 2.3 An example of
set-valued data

Individual	Items
1	I_1, I_2
2	I_1, I_3
3	I_1, I_2, I_3, I_4
4	I_1
5	I_2, I_4
6	I_3, I_4

To tackle the above two challenges in large-size set-valued data set, a differentially-private sanitization algorithm is proposed in [17] to protect set-valued data publishing by recursively partitioning a set-valued data set based on a context-free taxonomy tree. First, a context-free taxonomy tree is constructed. Then, all records in the data set are generalized to a single partition with the same generalized value. For example, all records in Table 2.3 can be generalized to a partition with generalized value $I_{\{1,2,3,4\}}$, which is the root in the taxonomy tree. Then, a partition is recursively cut into a set of disjoint sub-partitions with more specific generalized values. The partitioning continues recursively in a top-down manner and stops when the partition is considered to be "unpartitionable", that is, no further partitioning can be applied on the partition. To ensure differential privacy, the "unpartitionability" is determined in a noisy way. Finally, for each unpartitionable partition, the number of records in the partition is counted, and random Laplace noise is added to the count.

To achieve differential privacy, the partitioning procedure cannot be deterministic. Probabilistic operations are required to perform partitioning. Specifically, noise is added in the determination of unpartitionability. Since there are a sequence of operations, a certain portion of privacy budget ϵ is required to obtain the noisy size added to each operation. Thus, privacy budget allocation is required to carefully allocate the total privacy budget ϵ to each probabilistic operation to avoid unexpected algorithm termination. A naive allocation scheme is to bound the

Fig. 2.10 A context-free taxonomy tree

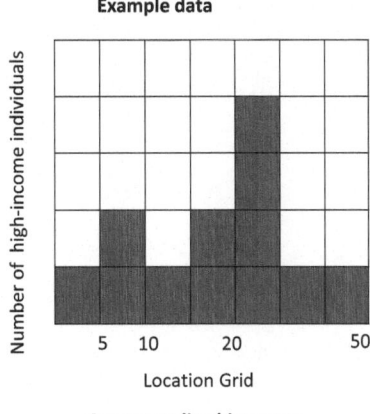

ID	Location Grid	Income
1	17	High
2	6	High
3	7	High
4	11	Low
5	18	High
6	16	High
7	1	High
...	...	...

Example data

Corresponding histogram

maximum number of partitioning operations and assign an equal portion to each operation. A more sophisticated adaptive scheme is proposed in [17] and it shows better results. The adaptive scheme reserves half portion of ϵ to add noise to counts, and the rest $\epsilon/2$ budget is used to guide partitioning. For each partitioning operation, the maximum number of partitioning operation needed in the future is estimated. And then based on the estimated number, a privacy budget is assigned. The portion of privacy budget assigned to a partitioning operation is further allocated to its sub-partitions determine the unpartitionability. Since all sub-partitions from the same partition operation contain disjoint individuals, the privacy budget portion used on each sub-partition can be the same and doesn't add up according to the parallel composition theory.

2.3.3 Applying Differentially-Private to Histogram Data

A histogram is a representation of tabulated frequencies, which are drawn as adjacent rectangles or bins with an area or height proportional to the frequency of the individuals in the bin. Histogram is usually used in database systems as an effective way to summarize statistical information in numerical domains. Figure 2.10 draws an example of histogram and corresponding data.

Formally, for a certain series of counts $D = \{x_1, x_2, \ldots, x_n\}$, $\forall x_i \mathbb{R}^+$, a k-bin histogram merges neighboring counts into k groups, i.e., a k-bin histogram can be presented as $H = \{B_1, B_2, \ldots, B_k\}$ where each bin $\{B_i\}$ covers an interval $[l_i, r_i]$ and a count $c_i \in [1, n]$ is associated with B_i. c_i is defined as total number of records falling into the interval $[l_i, r_i]$. In a valid histogram, all bins should cover the whole interval without any overlap between them.

The error for a certain bin B_i is defined by:

$$Error(B_i, D_i) \triangleq |c_i - \sum_{x_j \in [l_i, r_i]} x_j|^2. \tag{2.20}$$

Based on (2.20), the *Sum of Squared Error* (SSE) of histogram H is obtained by:

$$E\,(H, D) \triangleq \sum_{i=1}^{k} Error(B_i, D_i).$$

It is obvious that letting $c_i = \frac{\sum_{x_j \in [l_i, r_i]} x_j}{r_i - l_i + 1}$ for all i results in a minimal SSE $E\,(H, D)$. Histogram construction usually targets at finding the optimal histogram of a count sequence in terms of minimizing the SSE. It has been shown that the optimal histogram construction can be achieved via dynamic programming and costs $O(n^2 k)$. The dynamic programming problem can be formulated as

$$E(k, D) = \min_{B_k^i \in D} \left(E(k - 1, D \backslash B_k^i) + Error(B_k^i) \right), \tag{2.21}$$

where $E(k, D)$ is the minimal SSE for partitioning D into a k-bin histogram, B_k^i the ith sample output for the kth bin B_k, $E(k - 1, D \backslash B_k^i)$ the error for constructing partial histogram $H \backslash B_k^i$ with $k - 1$ bins.

Histogram publishing via differential privacy is investigated in [12, 49, 111]. Blum et al. [12] divides the input counts into bins with roughly the same counts to construct one-dimensional histogram. And then standard Laplace mechanism is applied on each count independently. According to Sect. 2.3.1, differential privacy is achieved with this simple method. However, this simple method overlooks a fact that the accuracy of a noisy histogram representation depends on its structure: a histogram with larger bins is usually more accurate than a histogram with smaller bins. This is because a histogram with larger bins needs smaller noise scale to satisfy the same level differential privacy.

By observing that the accuracy of a differential privacy compliant histogram depends heavily on its structure, Xu et al. [111] proposes two algorithms with different priorities for information loss and noise scales. The first algorithm consists of two steps: in the first step, Laplace noise is added to each original count x_i independently according to standard Laplace mechanism, where the noisy count sequence can be expressed as $\hat{D} = \{\hat{x}_1, \hat{x}_2, \ldots, \hat{x}_n\}$; in the second step, the optimal histogram based on noisy count sequence $\hat{D}$ is computed via dynamic

---

Algorithm 1 NoiseFirst Algorithm

Require: Count sequence D; bin number k; privacy parameters ϵ, count upper bound A;
Ensure: Noisy histogram $\hat{H}$;
1: **for** each x_i in a count sequence $D = \{x_1, x_2, \ldots, x_n\}$ **do**
2:    add independent Laplace noise with magnitude $1/\epsilon$;
3: **end for**
4: $\hat{D} \leftarrow \{\hat{x}_1, \hat{x}_2, \ldots, \hat{x}_n\}$;
5: Construct the optimal histogram based on $\hat{D}$ via dynamic programming (2.21);
6: Return the histogram $\hat{H} = \{\hat{B}_1, \hat{B}_2, \ldots, \hat{B}_k\}$ where $\hat{B}_i = \frac{\sum_{\hat{x}_j \in [l_i, r_i]} \hat{x}_j}{r_i - l_i + 1}$;

Algorithm 2 StructureFirst Algorithm

Require: Count sequence D; bin number K; privacy parameters ϵ, count upper bound A;
Ensure: Noisy histogram $\hat{H}$;
1: Construct the optimal histogram based on D via dynamic programming (2.21), and keep $Error(B_k^i)$, $E(k, D)$, for all $k \in \{1, \ldots, K - 1\}$, B_k^i, D;
2: $r_K \leftarrow n$;
3: **for** each k from $K - 1$ to 1 **do**
4:    **for** each possible kth bin $B_{k,i}$ **do**
5:       Compute the error $E(k, D, B_k^i) = E(k - 1, D \backslash B_k^i) + Error(B_k^i)$;
6:    **end for**
7:    Select $B_k \leftarrow B_k^i$ with probability $\propto \exp\left(-\frac{\epsilon_1 E(k, D, B_k^i))}{2K(2F+1)}\right)$;
8:    $D \leftarrow D \backslash B_k$;
9:    Add independent Laplace noise of magnitude $\epsilon_2(r_i - l_i + 1)$ to each count;
10:   Return the histogram $\hat{H} = \{\hat{B}_1, \hat{B}_2, \ldots, \hat{B}_k\}$;
11: **end for**

---

programming (2.21). This algorithm is referred to as NoiseFirst since noise is added before histogram construction. According to standard mechanism [12], the sensitivity of NoiseFirst is 1 and thus Laplace noise with magnitude of $1/\epsilon$ preserves ϵ-differential privacy on constructed histogram. Algorithm 1 summarizes this method.

Another counterpart algorithm is referred to as StructureFirst, where histogram construction comes before adding noise. The motivation for StructureFirst is that NoiseFirst does not leverage the reduced sensitivity after merging neighboring bins. To address this issue, StructureFirst construct histogram on the original data before adding noise to counts. Algorithm 2 summarizes the flow of StructureFirst. First, the optimal histogram is constructed based on the original count sequence D rather than noisy count sequence. The intermediate results in the process of dynamic programming are preserved to guide random bin merging. To add noise to the result histogram, StructureFirst adopts exponential mechanism to randomize the selection

ID	Location Grid	Income
1	17	High
2	6	High
3	7	High
4	11	Low
5	18	High
6	16	High
7	1	High
...	...	...

ID	Latitude	Longitude	Income
1	1	8	High
2	3	5	High
3	4	5	High
4	4	2	Low
5	4	4	High
6	6	1	High
7	2	3	High
...	...		...

Fig. 2.11 Contingency table

of bin's boundaries, where the probability of selecting a possible bin is proportional to the exponential function of the corresponding SSE $\exp\left(-\frac{\epsilon_1 E(k,D,B_k^i)}{2K(2F+1)}\right)$. The result histogram is proved to be ϵ-differentially-private.

The problem of releasing a set of consistent marginals of a contingency table is studied in [9] and [32], in which correlations among the marginals are extracted to reduce the additive noise. The marginals of contingency consists of multi-dimensional attributes and a sequence of counts. It can be seen as a multi-dimensional version of histogram. For example, the table in the upper side of Fig. 2.11 can be presented as a histogram, as shown in Fig. 2.10, while if we treat Income as attribute and count number of records falling into a two-dimensional (i.e., Location Grid and Income) region, it becomes marginals of the contingency table. The marginals can be high dimensional. As illustrated in Fig. 2.10, the Location Grid can be further de-composed as longitude and latitude and thus the resulting marginals becomes three-dimensional.

Count query on multiple dimensional tables can be interpreted as multiple cells. Each cell aggregates a subset of the rows in the table with corresponding counts, as shown in Fig. 2.12. The cells can be regarded as reflections of the original table

Fig. 2.12 Cell tables

Latitude	Income	Count
1-3	High	3
4-6	High	3
1-3	Low	0
4-6	Low	1

Cell {Latitude, Income}

Longitude	Income	Count
1-4	High	3
5-8	High	3
1-4	Low	1
5-8	Low	0

Cell {Longitude, Income}

Latitude	Longitude	Income	Count
1-3	1-4	High	1
1-3	1-4	Low	0
1-3	5-8	High	2
1-3	5-8	Low	0
4-6	1-4	High	2
4-6	1-4	Low	1
4-6	5-8	High	1
4-6	5-8	Low	0

Cell {Latitude, Longitude, Income}

on different subsets of dimensions with aggregated measures. However, a adversary can still infer sensitive information about an individual by just look at difference cells. To address this issue, instead of publishing exact counts, many proposals adopt differential privacy model to add random noise to the counts.

A simple way to release cell tables with differential privacy is directly applying Laplace mechanism. The first approach is to independently add Laplace noises with proper scales to each cell. The limitation of this approach lies in the fact that for a d-dimensional table there are 2^d cell tables in total. To satisfy ϵ-differential privacy, each cell needs to add Laplace noise $Lap(2^d/\epsilon)$, which grow exponentially with table dimensions. This significant amount of noise in a high dimensional table renders cell tables useless. Another issue is consistency among different cells. A key observation is that there exists correlations between cells, that is, lower-dimensional cells can be derived by merging some rows in higher-dimensional cells. If noises are added to each cell independently, the results may not be consistent across different dimensional cells. The inconsistency can be leveraged by an adversary to infer user's private data. For example, as all noises are independent, an adversary can compute the mean of multiple inconsistent counts to derive a more accurate results. Another issue of inconsistency is that data analyzers may interpret the inconsistencies as an evidence of bad data. A user study conducted by Microsoft [72] reported that only small inconsistencies in modified data were acceptable.

To address this issue, Ding et al. [32] divides the set of cell tables into two subsets. A subset of cell tables are selected and computed directly from the original data table. Standard Laplace noises are injected to these cell tables. This subset is called the initial subset. Then, based on the noisy initial subset, the count measure for the rest cell tables are computed. When the initial subset is larger, each of its cell tables requires more noises to preserve the same level of differential privacy, while on the other hand, the rest cell tables whose count measure is computed from the initial subset contain less noise. To choose a proper tradeoff that minimizes overall noise, a noise control framework is proposed in [32], in which two scenarios under the policy of Ministry of Health are discussed. In the first scenario, a set of cell tables are required to be released, and the objective is to minimize the maximum noise in the cell tables. The other scenario is the importance of each cell table is indicated by a weight function. The question is to decide which cell tables to release so that the noise level in each cell table is bounded by a pre-defined threshold. The objective is to maximize the sum of the weights of all released precise cell tables.

To achieve these goals, two optimization problems are formulated to select the initial subset in these two scenarios, respectively. These two optimization problems are proved to be NP-hard. The proof uses a non-trivial reduction from the *Vertex Cover problem* in degree-3 graph. A vertex cover of a graph G is a set S of vertices such that each edge of G is incident to at least one vertex in S. The Vertex Cover problem is to find the smallest vertex cover of a given graph. A brute-force approach for these problems is to enumerate all possible choices of the initial subset, which takes $O(2^d)$ time where d is the dimension of the original table. The brute-force approach is not practical for high-dimensional table. Approximation algorithms with polynomial time complexity are proposed: Bound Max Variance and Publish Most.

To cope with the first scenario, whose target is to minimize the maximum noise in cell tables, we can find the minimum noise threshold η such that there exists feasible solutions. Specifically, suppose we have algorithm for a subproblem $SubProb(S_i, n, \eta)$, which is described as: for a given threshold η and a positive integer n, is there an initial subset S_i with n cell tables that for all cell tables in S_i, the required noise is no larger than η? If the answer is yes, we set $SubProb(S_i, n, \eta) = YES$, otherwise $SubProb(S_i, n, \eta) = NO$. Then the problem can be interpreted as finding the minimum η such that $SubProb(S_i, n, \eta) = YES$. To find such η efficiently, binary search instead of brute-force search is used. The details are provided in Algorithm 3 and Algorithm 4. It is proved in [32] that the Bound Max Variance algorithm provides a logarithmic approximation and runs in polynomial time.

Consider the second scenario in which the objective is to maximize the sum of the weights of all released precise cell tables. Suppose the optimal solution for a

---

Algorithm 3 $SubProb(S_i, n, \eta)$

1: compute coverage $coverage(C)$ for each cell C;
2: $R \leftarrow \emptyset$;
3: $COV \leftarrow \emptyset$;
4: repeat the following steps n times:
5: pick the cell C' with max $|coverage(C') - COV|$;
6: add C' to R;
7: add $coverage(C')$ to COV;
8: **if** COV covers all cell tables **then**
9:    $S_i = R$;
10:    return YES;
11: **else**
12:    return NO;
13: **end if**

Algorithm 4 Bound Max Variance

1: $\eta_L \leftarrow 0$;
2: $\eta_R \leftarrow 2^{2d+1}/\epsilon^2$;
3: **while** $|\eta_R - \eta_L| > 1/\epsilon^2$ **do**
4:    $\eta \leftarrow \frac{\eta_R + \eta_L}{2}$;
5:    $\eta_R \leftarrow \eta$;
6:    **for** all $n = 1, 2, \ldots, 2^d$ **do**
7:       **if** there exists an n such that $SubProb(S_i, n, \eta) = YES$ **then**
8:          $\eta_L \leftarrow \eta$;
9:       **end if**
10:    **end for**
11:    return η_R;
12: **end while**

---

noise threshold η_o contains a set of n cell tables. It is proved that the problem is equivalent to finding a set of n cells subjected to the weighted sum of their covered cells is as high as possible. A greedy algorithm is considered to solve the problem for each choice of n. As described in Algorithm 5, each iteration finds the cell table that maximize the total weight of the cells who are not previously covered. This cell table and the corresponding newly covered cell tables are added to R and COV, respectively. At the end of the 2^d th iteration, the algorithm selects the best R over all possible n as the initial subset. It has been proven that Algorithm 5 provides $(1 - 1/e)$ approximation with running polynomial in 2^d.

There are several other studies on using differential privacy to release different types of data. As these works are tightly related to location data, this book does not elaborate them. Readers who are interested can refer to the follower papers. The brief ideas of the other state-of-the-arts are described as follows. *Privelet* [110] is proposed to lower the magnitude of additive noise to publish multi-dimensional

---

Algorithm 5 Publish Most

1: **for** $n = 1, 2, \ldots, |2^d|$ **do**
2:     **for** each cell C **do**
3:         compute coverage $coverage(C)$ based on η_o, n;
4:     **end for**
5:     $R \leftarrow \emptyset$;
6:     $COV \leftarrow \emptyset$;
7:     repeat the following steps n times:
8:         select a cell table C' with maximal $coverage(C') - COV$;
9:         add C' to R;
10:         add $coverage(C')$ to COV;
11: **end for**
12: return the best R_s

---

frequency matrix. *DiffGen* [75] connects generalization technique with differential privacy to releases an sanitized data table for classification analysis. DiffGen randomizes the generalization operation and adds noise to count in each partition. Recently, several works [30, 32, 80] try to handle multi-dimension issue in differential privacy. Peng et al. [80] and Cormode et al. [30] introduce differential private indices to reduce errors on multi-dimensional data sets.

2.4 Private Information Retrieval

The aim of private information retrieval (PIR) is to enable a user to retrieve an item from a database without revealing which item is retrieved. Typically, PIR is used in database query, which is related to the database-driven CRNs in which registered SUs need to query available channels from a database. A simple way to achieve this is for the user to query the whole database. Apparently, it is very inefficient and requires large amount of overhead. A natural question arises: can we use less communication overhead while still ensuring the privacy of the retrieval? PIR replies this question with a positive answer. A tutorial on PIR can be found in [114].

Let's describe the problem in a more concrete manner. A database can be modeled as [20]

1. a vector of bits $\mathbf{x} = [x_1, x_2, \ldots, x_n]$,
2. and a computational agent can do computations over $\mathbf{x}$ based on a query q.

Based on this database model, the problem of PIR can be stated as:

1. a user retrieves x_i,
2. i is unknown to the database.

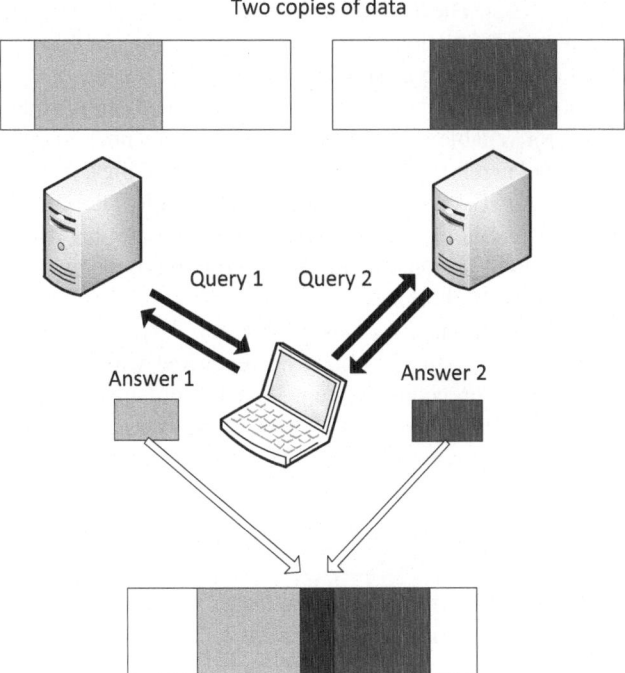

Two copies of data

Fig. 2.13 PIR: two databases illustration

Actually, the problem can be more strict: the database cannot know that $i \neq j$. This can be achieved by querying total n bits. Then the question here becomes can the user retrieve x_i with complete privacy while using less bits of communication? It is usually assumed that the user's PIR algorithm is known to the database. In such case, the answers are divided. Here we only outline the major PIR techniques. A detailed survey on PIR can be found in [41,78]. In [41], the authors conclude several cases:

- If the user adopts a deterministic algorithm, n bits communication is required.
- If the copy of the data is stored only in one database, and the database's computational power is unlimited, then n bits communication is required.
- If the copy of the data is stored only in multiple non-communicating databases, and the database's computational power is unlimited, then less than n bits communication is possible.

Figure 2.13 illustrates a two databases case. The user queries two databases separately, each of which stores a copy of the identical data. Each individual query carries no information about what the user wants to retrieve. The retrieval data is obtained by combining the two answers. The most efficient two-database PIR protocols known today require communication overhead of $\mathcal{O}(n^{1/3})$ [20]. Moreover, PIR techniques for three or more databases have been widely developed [7, 10, 35, 112]. The lower bounds and upper bounds have been discussed in many research works, but closing the gap between upper and lower bounds for PIR communication complexity is still an open problem.

The core technique adopted to design PIR is error-correcting codes, which are initially designed for reliable transmission over noisy channel, or reliable storage on partially corrupted medium. A category of error-correcting codes called Locally decodable codes or LDCs are leveraged to enable PIR. The core functionality of LDCs is to provide efficient random-access retrieval and high noise resilience simultaneously by reliably reconstructing an arbitrary bit of the message based on a small number of randomly chosen codeword bits.

A (k, δ, ε)-LDC encodes an n-bit message $\mathbf{x}$ to an N-bit codewords $\mathbf{C}(\mathbf{x})$, where for each $1 \leq i \leq n$, the ith bit x_i can be recovered with probability $1Ce$ by a randomized decoding procedure reading k codeword bits, even after $\mathbf{C}(\mathbf{x})$ is corrupted in at most δN bits. PIR and LDCs are strongly related. Short LDCs yield efficient PIR schemes and vice versa. In the following part, we describe how to derive a k-database PIR scheme from any perfectly smooth k-query LDC.

- First, each of the k databases $D_1, \ldots, D_k$ encodes a n-bit data $\mathbf{x}$ with a perfectly smooth LDC $\mathbf{C}(\cdot)$.
- Without loss of generality,we assume that a user wants to retrieve the ith bit x_i. Then, the user tosses random coins to generate k random queries $q_1, \ldots, q_k$, such that x_i can be computed from these k queries.
- The user sends k queries $q_1, \ldots, q_k$ to corresponding databases $D_1, \ldots, D_k$.
- Each database $D_j, j \in \{1, \ldots, k\}$ answers the query q_j with one bit $\mathbf{C}(\mathbf{x})_{q_j}$.
- The user reconstruct x_i by combining $\mathbf{C}(\mathbf{x})_{q_1}, \ldots, \mathbf{C}(\mathbf{x})_{q_k}$.

The above protocol preserves the privacy of x_i in that each query q_j is uniformly distributed over the set of codeword coordinates and thus the database D_j learns nothing from q_j. The total communication cost is $k(\log N + 1)$. As for surveys on information-theoretic PIR, you can refer to [97, 113].

Chapter 3
Location Privacy Preservation in Collaborative Spectrum Sensing

With the proliferation of mobile devices and the rapid growing of wireless services, cognitive radio networks (CRNs) have been recognized as a promising technology to alleviate the spectrum scarcity problem. The CRNs allow secondary users (SUs) to utilize the idle spectrum unoccupied by primary users (PUs). A major technical challenge in the CRNs is to acquire knowledge about spectrum occupancy properties through spectrum sensing. Recent standard proposals for CRNs (e.g., IEEE 802.22 WRAN [3], CogNeA [2]) adopt *collaborative sensing* to improve spectrum sensing performance, that is, the sensing data from multiple SUs is aggregated to learn the spectrum occupancy. On the other hand, in a realistic CRN setting, multiple service providers (SPs) operate on the same set of frequency bands in one geographic area, where each SP serves a group of SUs. This multi-SP scenario has been intensively discussed in the CRN literature (e.g., [29, 77]). Existing collaborative sensing schemes [18, 59, 61, 74, 82] also show that the performance of spectrum sensing can be improved when more SUs are involved in the collaboration since the spatial diversity can be better exploited with larger amounts of sensing data. Thus, there is strong motivation for multiple SPs to acquire the spectrum occupancy status collaboratively.

Although the collaboration among multiple SPs results in better sensing performance, it suffers from privacy threats that compromise SU's incentives to join the collaborative sensing. The experiments conducted in [64] demonstrate that malicious SPs or SUs (called adversaries) can geo-locate an SU based on its sensing data using localization techniques. It is also shown in [64] that an SU's sensing data can be recovered from the aggregated results if no privacy protection method is adopted. Thus, the location privacy of an SU can be compromised if its sensing data is known or inferred by an adversary. Being aware of the potential privacy threats, SUs may not want to participate in the collaborative sensing if their privacy is not guaranteed. Therefore, it is essential to guarantee the privacy of each SU's sensing data in collaborative sensing.

W. Wang and Q. Zhang, *Location Privacy Preservation in Cognitive Radio Networks*, SpringerBriefs in Computer Science, DOI 10.1007/978-3-319-01943-7_3,

However, most previous approaches on collaborative sensing have focused on spectrum sensing performance improvements [18, 59, 61, 74, 82] or security related issues [19, 62, 73], while privacy issues are less discussed. In this chapter, we will discuss the location privacy issues in collaborative sensing scenario. Li et al. [64] proposes two privacy preserving protocols to protect SU's sensing data from an untrusted server. We will describe these privacy protocols first. Then, we extend the privacy issues in the multi-SP collaboration context. We will discuss potential privacy threats and possible solutions.

To launch location privacy attacks, the adversaries can adopt certain RSS-based localization technique. Specifically, the adversary overheads the SU's RSS of different channels, and uses radio fingerprint to match the SU's RSS with physical location. In the experiments conducted by Li et al. [64], the RSS and physical location are highly correlated. Thus, to protect the SU's location privacy, it is essential to keep the sensing report private.

3.1 Modeling Collaborative Spectrum Sensing

Location privacy issues in collaborative spectrum sensing are divided into two contexts: single-service-provider context and multi-service-provider context. In this section, we give a brief overview of the collaborative spectrum sensing systems under these two different contexts. To give a comprehensive understanding of these two scenarios, we describe formalized models of SUs, PUs, FCs, and spectrum sensing in collaborative spectrum sensing.

In single-SP context, we consider a CRN as illustrated in Fig. 3.1, where a set of SUs $\mathscr{U}$ served by one SP collaboratively senses channels to learn the channel occupancy properties. Each SU $u \in \mathscr{U}$ senses a set of licensed channels $\mathscr{H} = \{1, \ldots, h, \ldots, H\}$, which may be dynamically used by PUs, and obtains a vector of normalized *received signal strength* (RSS) $\mathbf{r}_u = [r_{u1}, \ldots, r_{uh} \ldots, r_{uH}]$, where $\forall r_{uh} \in \mathbf{r}_u, 0 \leq r_{uh} \leq 1$ with 1 for the strongest signal and 0 for no signal. Then, the SU u sends the vector $\mathbf{r}_u$ to the FC run by the SP as the sensing report.

At each period, the FC combines the sensing reports from all SUs to derive an aggregated result. Without loss of generality, the aggregation can be modeled by the most commonly used EGC, as discussed in Chap. 1. In the EGC model, a simple summation of collected signal strengths is used as the final decision statistic.

In multi-SP context, we consider a CRN as illustrated in Fig. 3.2, where a set of SUs $\mathscr{U} = \bigcup_{p=1}^{P} \mathscr{U}^{(p)}$ served by a set of SPs $\mathscr{P} = \{1, \ldots, p, \ldots, P\}$ collaboratively senses channels to learn the channel occupancy properties. Each SU $u \in \mathscr{U}$ senses a set of licensed channels $\mathscr{H} = \{1, \ldots, h, \ldots, H\}$, which may be dynamically used by PUs, and obtains a vector of normalized RSS $\mathbf{r}_u = [r_{u1}, \ldots, r_{uh} \ldots, r_{uH}]$, where $\forall r_{uh} \in \mathbf{r}_u, 0 \leq r_{uh} \leq 1$ with 1 for the strongest signal and 0 for no signal. Then, the SU $u \in \mathscr{U}^{(p)}$ sends the vector $\mathbf{r}_u$ to its SP p as the sensing report.

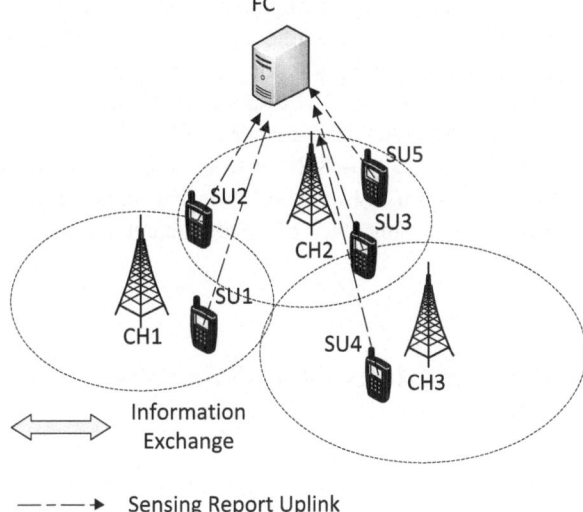

Fig. 3.1 System model for collaborative spectrum sensing with one FC. In this example, five SUs are served by one FC, and sense three channels, i.e., CH1, CH2 and CH3. Each SU sends sensing reports containing RSS values in the three channels to the FC. The FC combines the sensing reports to learn the spectrum

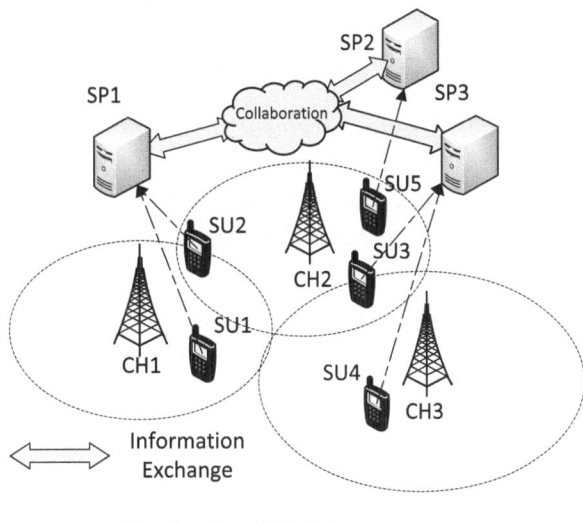

Fig. 3.2 System model for collaborative spectrum sensing with multi-SP. In this example, five SUs are served by three SPs, and sense three channels, i.e., CH1, CH2 and CH3. Each SU sends sensing reports containing RSS values in the three channels to its own SP. The three SPs exchange information with each other to collaboratively learn the spectrum

Multiple SPs $\mathscr{P}$ collaboratively learn channel availabilities by considering the sensing reports from all SUs $\mathscr{U}$ without revealing the sensing data of any individual SU. We make no assumptions about the aggregation functions or learning techniques of the collaborative sensing. The targets of collaboration among SPs can be manifold. SPs can aim to detect the presence of PUs on a channel using certain detection algorithms (e.g., [18, 59, 74, 82]), or learn the spectrum occupancy properties across multiple channels using some machine learning techniques (e.g., [61]). At the end of collaborative sensing, each SP sends the detection or learning results all its SUs as feedback.

3.2 Location Privacy Attacks in Collaborative Spectrum Sensing

Last section provides an overview of the collaborative spectrum sensing under the contexts of single SP and multiple SPs. We see that the major difference between these contexts is the interaction between SPs. Concretely, in single-SP scenario, there is only one SP, while in multi-SP scenario, there exists information exchange between multiple SPs. This difference leads to different privacy threats in these two scenarios. In this section, we elaborate possible privacy attacks under these two settings.

3.2.1 Attacks Under Single-Service-Provider Context

Since the sensing results are highly correlated to user's physical location, which can be exploited by adversaries to launch location privacy attacks. Several privacy attacks are reported in [64]:

- *Internal CR Report-Location Correlation Attack.* An untrusted FC can infer SU's location by looking at the SU's sensing reports. A malicious SU can also overhear other SUs' sensing reports by eavesdropping and infer their physical location based on correlation between report and location.
- *External CR Report-Location Correlation Attack.* The openness of wireless communication enables an external adversary to eavesdrop SU's sensing report and then to link physical location with sensing report.
- *Internal Differential CR Report-Location Correlation Attack.* Similar to privacy attacks considered by differential privacy, this attack is launched by exploiting the difference in the aggregated results. Typically, the FC can be the adversary of this attack. The adversary first records the aggregated sensing results before and after an SU joins or leaves the collaboration, and then estimates the SU's location by measuring the difference in the sensing results.

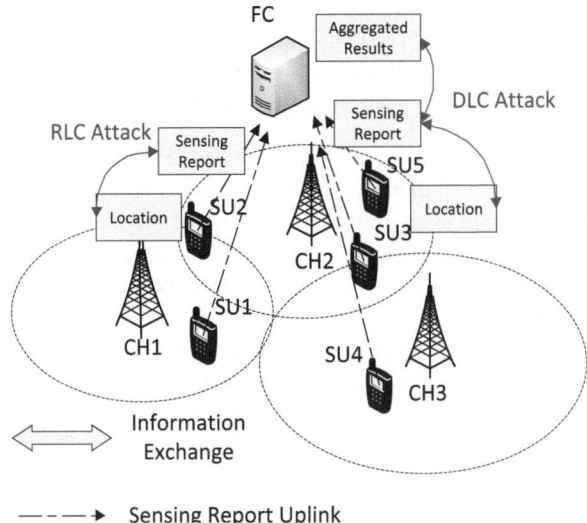

Fig. 3.3 RLC and DLC attacks in single-SP collaborative spectrum sensing scenario

The first two attacks, referred to as RLC attack, are basically the same in the privacy preservation perspective. RLC attacks are launched when SUs report their sensing results to the FC. While on the other hand, the third attack, referred to as DLC attack, occurs at the FC. Figure 3.3 depicts where RLC and DLC attacks could occur in a single-SP collaborative spectrum sensing scenario.

The adversaries can launch RLC attacks by simply adopting a RSS-based localization technique to link the sensing reports to physical locations. As for DLC attacks, the adversaries can first derive an SU's sensing report by comparing the changes of the aggregated results during the SU's departure or arrival.

3.2.2 Attacks Under Multi-Service-Provider Context

Multiple SPs $\mathscr{P}$ collaboratively learn channel availabilities by considering the sensing reports from all SUs $\mathscr{U}$ without revealing the sensing data of any individual SU. We make no assumptions about the aggregation functions or learning techniques of the collaborative sensing. The targets of collaboration among SPs can be manifold. SPs can aim to detect the presence of PUs on a channel using certain detection algorithms (e.g., [18,59,74,82]), or learn the spectrum occupancy properties across multiple channels using some machine learning techniques (e.g., [61]). At the end of collaborative sensing, each SP sends the detection or learning results all its SUs as feedback.

We assume that an SU only trusts its own SP, and does not want its sensing data to be inferred by other SPs or SUs, which are considered as adversaries. The adversaries are assumed to be *semi-honest* [5]. Semi-honest adversaries correctly follow the collaborative sensing scheme, yet attempts to learn the sensing data of the target SU by analyzing the information received in the collaboration. Specifically, there are three kinds of attacks considered in this paper.

- *SP Attack.* An SP adversary has perfect knowledge about the sensing data of all its SUs and collects information offered by other SPs during the collaboration. The SP adversary tries to infer the sensing data of the SUs served by other SPs based on its knowledge and the information it has attained in the collaboration. Multiple SPs may collude by combining their knowledge and information to infer other SUs' sensing data.
- *SU Attack.* One or multiple SU adversaries try to infer the sensing data of other SUs based on the feedback results and their own sensing data.
- *SP-SU Collusion Attack.* SP adversaries and SU adversaries could further collude by combining their knowledge and information to launch an attack.

Note that for an SU $u_i^{(p)} \in \mathcal{U}^{(p)}$, even other SUs served by the same SP p are not trustworthy. Thus, in the worst case, all SPs and SUs collude except the target SU and its SP.

3.3   Privacy Preserving Spectrum Sensing

In previous sections, we elaborated the single-SP and the multi-SP collaborative spectrum sensing scenarios and possible privacy threats. In this section, we discuss how to protect SU's privacy while still preserving the accuracy of collaborative spectrum sensing. In particular, we describe existing privacy preservation approaches as well as other possible solutions.

3.3.1   Privacy Preservation Under Single-Service-Provider Context

Li et al. [64] proposes a framework to cope with privacy threats in the single-SP collaborative spectrum sensing scenario. The framework consists of two protocols to deal with RLC and DLC attacks respectively.

To prevent RLC attacks, the framework adopts secret sharing technique to enable the FC to obtain the aggregated results without knowing each individual sensing report. The core idea is each SU encrypts its sensing report with its secret and the FC cannot decrypt the secret unless all sensing reports are collected. The detailed protocol is described as follows:

- *Key generation.* A trust third party generates a serial secret keys $\{s_1, \ldots, s_n\}$ that each key for an SU and a key for the FC s_{FC}, where the secret keys satisfy the condition $\sum_{i=1}^{n} s_i = -s_{FC}$.
- *Sensing report encryption.* Then each SU u_i performs spectrum sensing on channel h and encrypts its sensing report $r_{i,h}$ as

$$c_{i,h} = g^{r_{i,h}} H^{s_i}, \tag{3.1}$$

where H is a hash function modeled as a random oracle. The SU sends $c_{i,h}$ instead of $r_{i,h}$ to the FC.

- *Aggregation.* The FC aggregates sensing reports from all SUs by

$$A = H^{s_{FC}} \prod_i c_{i,h}. \tag{3.2}$$

We can see that $\prod_i c_{i,h} = g^{\sum_i r_{i,h}} H^{-s_{FC}}$ and thus $A = g^{\sum_i r_{i,h}}$. The summation of sensing results can be derived by computing the log of A base g. Since $\{s_i\}$ is unknown to the FC, the FC is unable to recover the sensing results $\{r_{ih}\}$.

However, the major limitations of this protocol are listed as follows:

- It requires all sensing reports to decode the aggregated result. This makes the protocol not quite practical. Since in real collaborative spectrum sensing setting, SUs can join or leave the collaboration. It would be unable to cope with the dynamics in collaboration. Moreover, the wireless channel for sensing reports may be unreliable, which makes some of the sensing reports not accessible by the FC.
- Aggregation function is limited to summation, which is not the case for most aggregation scheme. As the secret sharing technique leveraged in this protocol is designed for summation function, it cannot be used to recover aggregated sensing result if the aggregation function is not summation, which is the case for many collaborative spectrum sensing techniques.
- Overhead. The secret sharing necessarily involves extra communication cost in the collaborative spectrum sensing, which will result in extra delay and computational cost.

To cope with DLC attack, another protocol is proposed. The core idea is as follows. when an SU joins or leaves the collaboration, other SUs sends dummy reports to the FC instead of real sensing reports with a pre-defined probability. The dummy report can be the sensing result of the FC. This approach does not introduce extra noises since the dummy report is also a real sensing result. Instead, it only increases the weight of sensing report rom the FC. The protocol is described in Algorithm 6.

Algorithm 6 Dummy Report Protocol

1: **if** there is an SU joining or leaving the collaboration **then**
2:      **for** each SU $u_i \in \mathscr{U}$ **do**
3:          **for** subsequent time slots **do**
4:              **if** the sensing report $\mathbf{r}_i$ doesn't change **then**
5:                  Submit the sensing report $\mathbf{r}_i$ with probability δ, and submit the dummy report $\mathbf{r}_0$ with probability $1 - \delta$;
6:              **end if**
7:          **end for**
8:      **end for**
9: **end if**

3.3.2 Privacy Preservation Under Multi-Service-Provider Context

Privacy preservation is also studied in distributed settings, in which the aggregated results are derived from multiple partitions of data held by different entities. This type of privacy preservation is related to privacy preserving spectrum sensing in multi-SP scenario. While these entities do not trust each other and only share limited information about their data to contribute in the aggregation. The distribution settings are classified into *vertical partitioning* and *horizontal partitioning*. In the vertical partitioning setting [99, 100], each entity holds different attributes of the same set of individuals. In the horizontal partitioning setting [115, 116], each entity holds the same set of attributes of different individuals. The multiple SP setting in this paper falls into this category. Yu et al. [116] consider a clinic setting where patients' records are horizontally distributed among multiple hospitals, and proposes a privacy-preserving Cox regression model. To protect privacy, the original data is projected to a lower dimension within each hospital, and only computation results on the projected data are shared to learn the Cox model. However, the achieved privacy is not quantified by Yu et al. [116]. In [115], a cryptographic technique called *secure multiparty computation* is leveraged to allow multiple data holders to privately learn a supporting vector machines (SVM) classifier. Nonetheless, all these methods fall short under the collusion attacks in multi-SP context. Thus, more strict privacy preservation schemes that are specially designed for multi-SP collaborative spectrum sensing are required, which currently is still an open issue.

Chapter 4
Location Privacy Preservation in Database-Driven Cognitive Radio Networks

Besides spectrum sensing, geo-location database query approach is another typical approach to obtain spectrum availabilities at SU's location. The database query approach is enforced by the latest FCC's rule [28] released in May 2012. In this chapter, we will describe a general model for the database-driven cognitive radio networks. Then, we discuss potential privacy threats in such model. Different adversary models are discussed, and corresponding defending approaches are presented.

4.1 Location Privacy Attacks in Database-Driven Cognitive Radio Networks

A typical database-driven CRNs consist of a set of PUs (e.g., TV transmitters/receivers, wireless microphones in TV white space), a set of SUs, and a database, as shown in Fig. 4.1. An SU can either be a wireless transmission link or a wireless access point serving multiple end users. FCC requires the SUs to periodically access the database. In each period, before accessing the channels, the SUs should connect the database to obtain a list of available channels.

4.1.1 Potential Privacy Threats When the Database is the Adversary

The database is assumed to be semi-honest, that is, the database exactly follows the protocol but tries to infer SU's locations. In [40], the knowledge of database is assumed to include the complete communication content between SU and the database, and the spectrum utilization information of SUs. It is also assumed in [40] that the database has unlimited computational power.

W. Wang and Q. Zhang, *Location Privacy Preservation in Cognitive Radio Networks*,
SpringerBriefs in Computer Science, DOI 10.1007/978-3-319-01943-7__4,
© The Author(s) 2014

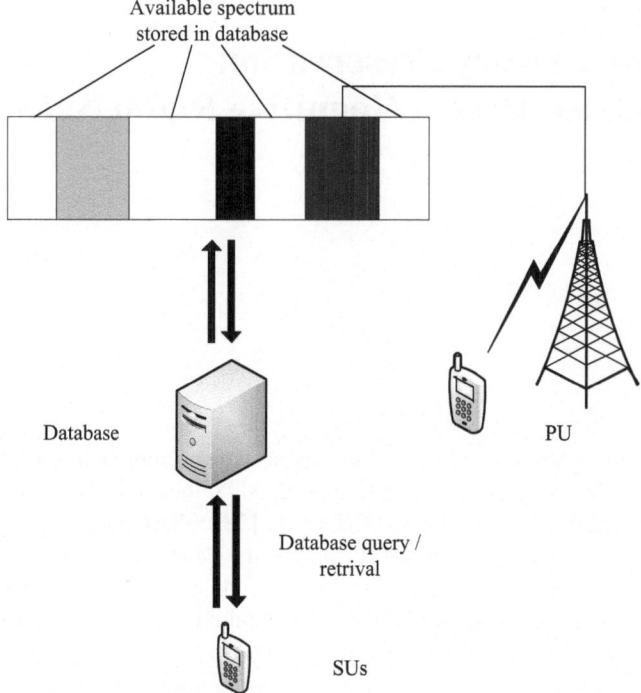

Fig. 4.1 System model for database-driven CRNs

Gao et al. [40] discovered that the database can successfully obtain the location of an SU based on the SU's channel usage pattern. A basic observation is that an SU can use a channel only if the SU is located outside the signal coverage of the PUs that allocate on that channel. This observation is straightforward since otherwise the SU would cause interference to the PU's transmissions, which is not allowed in the CRNs. Another observation is that the SU has to dynamically use the channels to avoid interference to the PUs. Combining the two observations, the SU's location can be narrowed down to the intersection area of the complements of PU's coverage over time.

Figure 4.2 illustrates how to perform such an attack. The unavailable area of the channel is the coverage of the PU using this channel, while the available area is the complement of the PU's coverage. t_1, t_2, t_3 denotes three different time slots, and in each time slot a PU with different location transmits data on that channel. We can see in this example, the available areas of the channel vary over the three time slots due to the location changes of the active PU. Assuming that the location of the SU doesn't change over time, the adversary can perform an intersection attack, which is referred to as complement coverage attack, by intersecting the available areas over three time slots to narrow down the location of the SU. It can be seen in the example that by combining the available areas in the three time slots, the location grid of the SU can be identified.

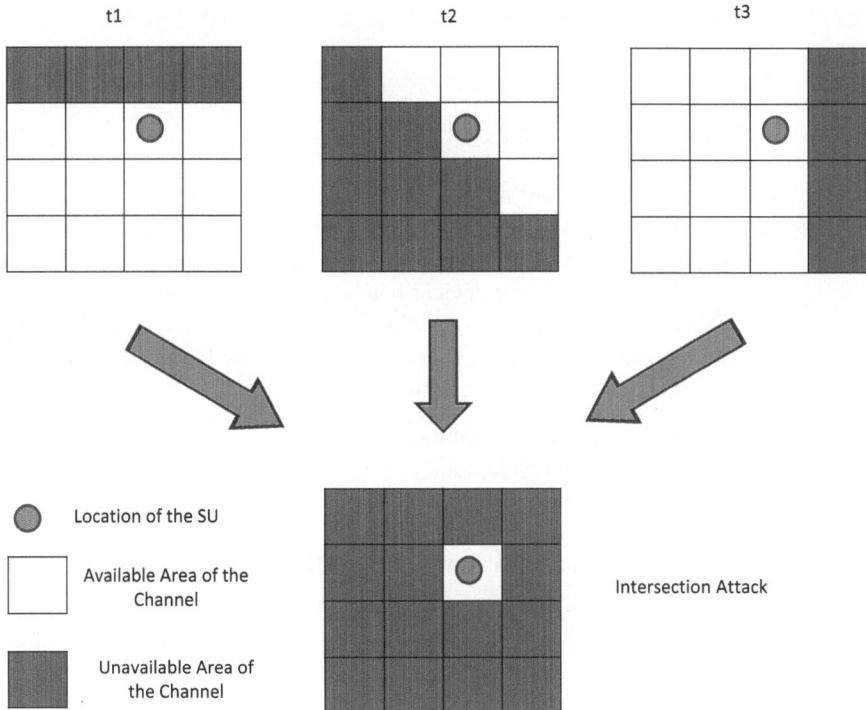

Fig. 4.2 An example of coverage complement attack

Experiments are set up in [40] to validate such an attack. In the experiments, the TV white space database is setup by storing the spectrum information in Los Angeles (LA) provided by TVFool [1] and implementing with FCC's regulations. A total availability information of 29 channels in an area of 75 km × 75 km is stored in the database. The area is divided into 100 × 100 grids. They ran Monte Carlo simulations and the results show that in most cases, the SUs can be localized within five grids.

However, this complement coverage attack has several limitations:

- The SU is assumed to be static over time. This is the basic assumption adopted in complement coverage attack. However, in real settings, the SUs are usually mobile over time. If the SU is static, there are many ways to localize the SU, e.g. using RSS-based localization techniques.
- The adversary knows the SU's channel usage information but does not know the SU's query. Otherwise, there is no need to perform such an attack since the adversary can extract the location directly from the query. So the adversary cannot be the database. While without the information of database, an external adversary can hardly know the coverage of the PUs. Thus, the adversary that can perform complement coverage attack is somehow not quite intuitive.

4.1.2   Potential Privacy Threats When A Secondary User is the Adversary

Let's first review the geo-location capability in database-driven TV white space CRNs required by FCC and OFCOM.

The SU devices are divided into three types: Mode I device, Mode II device, and fixed device. Fixed device refers to the access point or base station whose location is fixed. Fixed device needs to register its location to access white space. While Mode I and Mode II devices can be mobile devices whose locations can change over time. FCC and OFCOM have policies on these devices to make sure their usage of the white space causing no harm to the TV usage. AS for Mode II (master device in OFCOM's regulation) devices, they are required to

- be capable of identifying its geo-location.
- be able to access database via internet.
- receive available channel/frequency information from the database.
- be able to control its client devices.
- access database at least once a day (every 2 h by OFCOM).

As for Mode I devices, they are not required to have the above mentioned capabilities. The only requirement on Mode I devices is that they must be under the control of Mode II devices or fixed devices.

The geo-location capabilities of Mode II devices required by FCC are listed as follows.

- localization capability with 50 m of resolution.
- check its location every 60 s.
- access to the database after moving beyond 100 m.

While OFCOM has similar but more strict requirements:

- with the accuracy of the location with 95 % certainty. This accuracy should reflect the maximum area of operation within which slave devices can be located.
- receiving from the database as set of parameters including frequencies of allowed operation, associated power levels, geographic validity of operation, time validity of operation.
- ceasing transmission immediately where the time validity expires or where it moves outside of the geographic area of validity.

We can see that FCC and OFCOM have very detailed requirements for Mode II device's geo-location abilities, which have direct impact on the feedback from the database. Let's think reversely: what if a Mode II device doesn't submit a correct location, as depicted in Fig. 4.3? Apparently, due to the openness of the software defined radio, it is very easy for an SU to modify its location data without being caught. Consider a more powerful malicious SU. Assume that the malicious SU knows neighboring SUs' locations by overhearing their signals and then applying localization algorithm. As such, the malicious SU can submit other's location as the falsified location. By doing this, the malicious SU can retrieve neighboring SUs available channels.

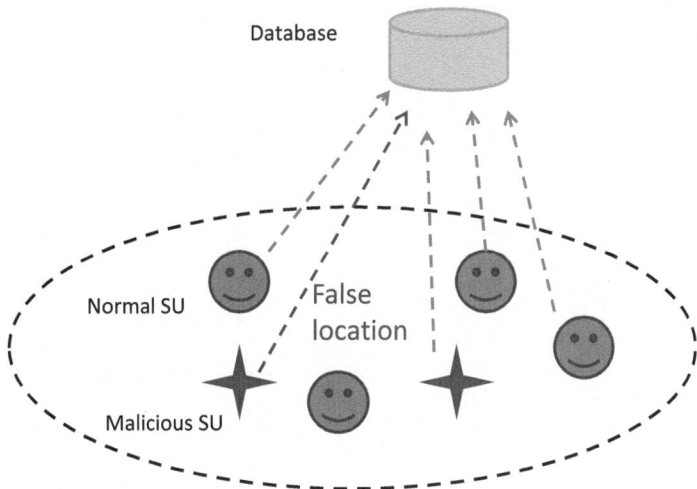

Fig. 4.3 System model for database-driven CRNs

The basic observation is that neighboring SUs have different available channels. WhiteFi [8] conducted some measurements in nine buildings on a campus spanning an area of approximately $0.9\,\text{km} \times 0.2\,\text{km}$. They find that median number of channels available at one point but unavailable at another is close to 7. The measurements conclude that spatial variation exists on smaller scales due to: (1) obstructions and construction material, and (2) wireless microphones which have typical transmission ranges of a few hundred meters ranging from small-scale lecture rooms to large-scale music and sporting events. On the other hand, since the frequency of white space is lower than WiFi, the communication ranges are larger in white space and is expected to exceed 1 km. Therefore, neighboring SUs usually have different available channels as well as shared channels.

Knowing neighboring SUs' available channels, the malicious SU would prevail over others when contending the white space channels. For example, the malicious SU can aggressively transmit data on the channels shared with neighbors to make others think that the channels are very busy. And thus, the neighboring SUs may allocate other channels. As such, the malicious SUs may exclusively use the shared channels.

4.2 Privacy Preserving Query for Database-Driven Cognitive Radio Networks

To thwart complement coverage attack, [40] proposes a scheme called Privacy Preserving Spectrum Retrieval and Utilization, or PriSpectrum for white space database. PriSpectrum consists of two parts: Private Spectrum Availability

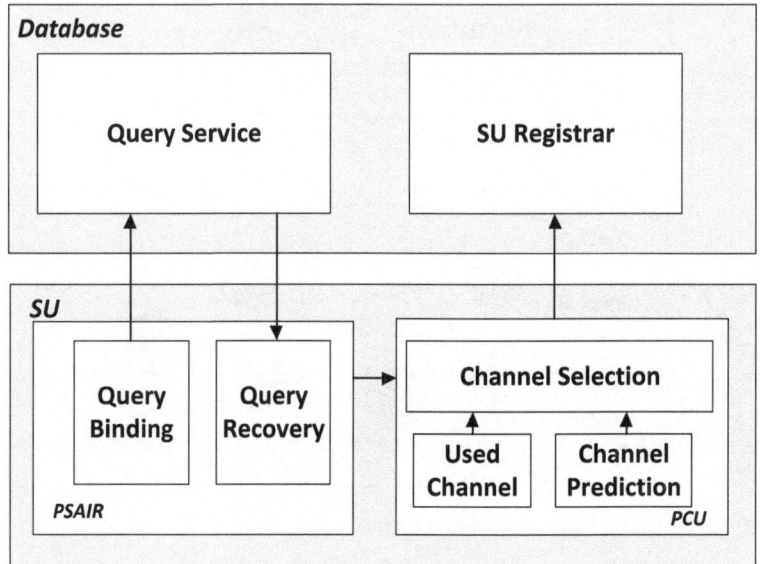

Fig. 4.4 PriSpectrum overview

Information Retrieval, i.e., PSAIR, protocol and Private Channel Utilization, i.e., PCU, protocol. PSAIR protocol provides privacy preservation for database query based by leveraging PIR technique, and PCU protocol provides private spectrum utilization.

Figure 4.4 depicts the overview of PriSpectrum. The basic idea of PSAIR is that to retrieve element a_{ij} without revealing the indices i, j to the database. To achieve this, the SU blinds two vectors with two blinding factors which can be removed by the SU. The scheme of blinding each vector is borrowed from [98], while PriSpectrum extends the scheme to two dimensions and further proves its correctness.

PSAIR only assures the privacy of location in the query, to achieve private channel utilization, PCU tries to reduce the location privacy leakage without changing existing access rules. The observation of designing PCU are as follows: (1) The location of the SU can be narrowed down only when the SU accesses to a channel with different PU coverage. (2) The SUs choosing a more stable channel will have less channel transitions. Based on these observations, PCU adopts two rules for channel utilization: first, the SU should choose the channels that have been used before with higher priority; second, the SU should choose the stable channels with higher priority. Therefore, when an SU needs to make a channel switch, it should first consider the most stable channel from the used channel list. If there is no available used channel, then the SU should choose the most stable channel. PCU protocol is summarized in Algorithm 7.

Algorithm 7 PCU Protocol

1: **if** A channel switch is needed **then**
2:      **if** there is an available channel in the used channel list **then**
3:          Choose the most stable one;
4:      **else**
5:          choose the most stable channel in the available channel list;
6:      **end if**
7:      Output the chosen channel;
8: **end if**

To measure the stability of a channel, the channel availability information at the SU's location is used. The availability information is modeled as a continuous Markov process with state transition rate observed from historical information. The channel h's state can be represented as

$$Q_h = \begin{pmatrix} -\lambda_h & \lambda_h \\ -\mu_h & \mu_h \end{pmatrix} \qquad (4.1)$$

where λ_h is the transition probability from being unavailable to being available, μ_h the transition probability from being available to being unavailable. Based on Q_h, the state transition matrix can be computed as

$$P^h(t) = \exp(Q_t) = \begin{pmatrix} -p_{00}(t) & p_{01}(t) \\ -p_{10}(t) & p_{11}(t) \end{pmatrix} \qquad (4.2)$$

where 1 stands for being available while 0 otherwise. $p_{11}, p_{10}, p_{01}, p_{00}$ can be computed from λ_h and μ_h.

Based on state transition matrix, the expectation of a channel's available duration can be obtained

$$\mathbf{E}[t_1] = tp_{11} + 2tp_{11}^2 + \ldots = \frac{tp_{11}}{(1 - p_{11})^2}. \qquad (4.3)$$

Thus, to choose the most stable channel, the SU needs to choose a channel with the largest p_{11}.

Chapter 5
Future Research Directions

Information sharing and transmission is an essential part of SUs in CRNs. Collaborative spectrum sensing requires the sharing of sensing reports between SUs and the FC. In database-driven CRNs, the location information and channel availability information are required to be transmitted between the database and registered SUs. Location privacy preservation is a promising approach to incentivize SUs to participate in the CRNs. The general form of privacy preservation is to transform the original sensitive data including sensing reports and location query into some anonymous from that are still useful and have limited impacts on the performance. Location privacy in CRNs is highly related to location privacy studied in participatory sensing and privacy preserving data publishing. However, due to the distinct features of CRNs, privacy preservation in CRNs has its own unique challenges. For example, the performance degradation caused by inaccurate location data in participatory sensing is allowable, while in collaborative spectrum sensing, the performance degradation means interference to PUs which is a very serious issue and is not allowed. Besides, the data formats, network protocols, aggregation rules are quit different in CRNs and other scenarios.

Location privacy preservation in CRNs still has many issues that have not been solved. Existing works studying location privacy issues in CRNs shows the openness of this direction by exploiting new privacy attacks in different CRN scenario. While some practical issues and scenarios are still unexplored. For instance, recently the FCC conditionally designated nine entities as TV white space device database administrators. How to protect privacy under such multiple SPs context has not been studied yet. In this chapter, we will discuss some possible research directions in the future.

W. Wang and Q. Zhang, *Location Privacy Preservation in Cognitive Radio Networks*,
SpringerBriefs in Computer Science, DOI 10.1007/978-3-319-01943-7_5,
© The Author(s) 2014

5.1 Distributed Cognitive Radio Networks

As mentioned in previous section, CRNs can be run by multiple SPs. Currently, the FCC conditionally designated nine entities as TV white space device database administrators. The conditional database administrators are Comsearch, Frequency Finder, Google, KB Enterprises LLC and LS Telcom, Key Bridge Global LLC, Neustar, Spectrum Bridge, Telcordia Technologies, and W5db LLC. These SPs will operate on the same TV white space and will be competitors. Thus, the schemes and protocols designed for the single-SP scenario can only prevent internal attacks, while across SP attacks may take very different attacking strategies. Some possible scenarios are listed below.

- Location privacy leakage impacts on network performance. Previous works focus on how to protect location privacy, which is rather a concern at individual level. However, the location privacy should be risen up to system performance level when there are multiple SPs co-located in a same geographic area. Since each SP tries to maximize their own performance and profits by offering better service to its users. To achieve this, other SP's private information, such as user distribution, channel allocation, etc., can be leveraged. Does the location information important? Will it cause network unfairness among users from different SPs? It would be interesting to investigate this direction in the future.
- Privacy leakage among multiple SPs. Previous work focus on how to protect user's privacy from other users or the SP, while in multi-SP setting, the privacy protection among SPs should be considered. Some existing work on distributed privacy preservation as well as database privacy is related. While the differences are quite obvious: Besides SPs, the actions the SUs make this problem more complicated. The SUs can switch between multiple SPs and thus can carry information in the switching.

5.2 Privacy Preservation Against More Intelligent Adversaries

Most existing work assumes that adversaries take fixed attacking strategies that will not change with time. This assumption is valid for offline attacks, e.g. analyzing user's personal information and preferences. However, some adversaries can launch real-time attacks in CRNs, such as location-based advertising and malicious attacks with criminal intent. In such cases, it is highly possible that the adversaries will adapt their attacking strategies according to the environmental dynamics and user's strategies. This kind of intelligent adversaries have not yet been considered in CRNs.

Unlike previous works where adversaries adopt a specific attack algorithm, a more powerful adversary model can be assumed, that the adversary changes its attacking strategies adaptively, and no assumption should be made on the attack

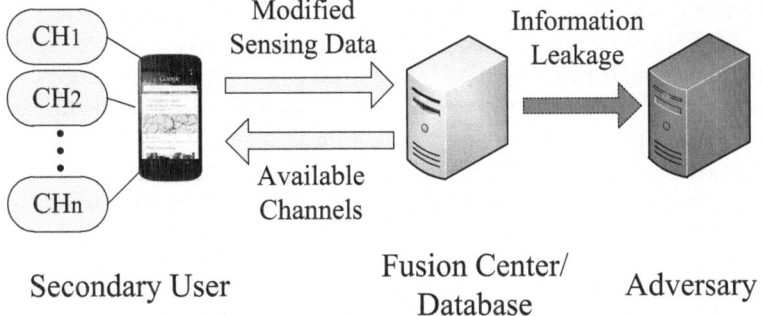

Fig. 5.1 Privacy threats from more intelligent adversaries

models, that is, the adversary can choose between different attack algorithms. To guarantee user's utility from all kinds of adversaries, the worst case assumption should be made: the adversary is *malicious* and aims at minimizing user's utility through a series strategic privacy attacks. Therefore, the SUs and the malicious adversary have opposite objectives. Moreover, since the environment (e.g., channel condition, channel availability) is considered to keep changing over time and both the SU and the adversary can make different actions at different times.

5.2.1 Modeling Privacy Threats

We describe a possible model of privacy issues in CRNs when intelligent adversaries adaptively adjust their strategies. Figure 5.1 illustrates a threat model where an SU senses multiple channels and submit the sensing results to an FC or a database to retrieve available channels. The SU applies a certain existing privacy-preserving technique (e.g., [48, 79, 104]) to modify the raw sensing data before releasing them to the applications. Based on the smartphone user's modified sensing data, the FC or database returns available channels to the SU and leak SU's sensing data to adversaries.

The SU senses multiple channels and releases the modified sensing data to the FC or database periodically according to the FCC's regulation, where a period is referred to as a time slot. It can also be assumed that the SU has installed a privacy-preserving middleware (e.g., MaskIt [46]) to control the released data granularity of each sensing record. The adversaries can only access user's released data that are modified by the privacy-preserving middleware, but the raw sensing data is kept private to the SU. The FC or the database makes decisions based on the modified sensing data. Normally, the released sensing data with coarser granularity leaks less information about the user, while the accuracies of channel availabilities computed by the FC or the database are also compromised. Due to SU's behaviors and activities, the SU's locations keep changing. It has been shown that human behaviors and activities extracted can be modeled well with a two-state Markov chain [55, 67]. Thus, we assume that the transitions between the SU's location contexts are modeled by a Markov chain. The adversary corrupts the FC or the database to access modified the sensing data, while cannot access the raw sensing data, since the privacy-preserving middleware is assumed to be non-corrupted. We also assume that the adversary can access the data of a subset of sensors in a time slot due to its computational constraint or limited bandwidth used for accessing data.

5.2.2   Modeling Interactions Between Users and Adversaries

An SU can encounter a set of location contexts $\mathscr{C} = \{c_1, \ldots, c_n\}$, where a subset of them are considered to be sensitive and disclosure of the sensitive contexts is considered to be harmful to the user. The privacy of contexts can be claimed by the user via special mobile phone applications (e.g., [96]). The SU's context privacy breaches if an adversary successfully infers that the SU is in its sensitive context.

The adversary is assumed to be able to obtain the reported sensing data at the time when the SU submits its sensing data. Note that an untrusted database or FC itself can be an adversary. Generally, it is assumed that the adversary knows the mobility model of an SU, and tries to compromise user's privacy by launching privacy attacks. Unlike previous works where adversaries adopt a specific attack algorithm, we assume a more powerful adversary model that the adversary changes its attacking strategies adaptively, and we make no assumption on the attack models, that is, the adversary can choose between different attack algorithms. To guarantee SU's utility from all kinds of adversaries, we make the worst case assumption: the adversary is *malicious* and aims at minimizing user's utility through a series strategic privacy attacks. Therefore, the SU and the malicious adversary have opposite objectives and thus their dynamic interactions can be modeled as a zero-sum game. If the adversary concerns about its cost of attacking, the adversary's utility is not exactly the opposite of the SU's utility, and thus the interactions between them can be modeled as a general sum game. Moreover, since both the SU and the adversary make different actions at different times, the SU and the adversary should consider their long-term utilities.

References

1. Tv fool. 2012.
2. CogNeA: Cognitive Networking Alliance. http://www.cognea.org.
3. IEEE 802.22 WRAN WG on Broadband Wireless Acess Standards. http://www.ieee802.org/22.
4. N.R. Adam and J.C. Worthmann. Security-control methods for statistical databases: a comparative study. *ACM Computing Surveys (CSUR)*, 21(4):515–556, 1989.
5. R. Agrawal and R. Srikant. Privacy-preserving data mining. In *Proc. ACM SIGMOD International Conference on Management of Data (SIGMOD)*, 2000.
6. H. Ahmadi, N. Pham, R. Ganti, T. Abdelzaher, S. Nath, and J. Han. Privacy-aware regression modeling of participatory sensing data. In *Proc. ACM Conference on Embedded Networked Sensor Systems (SenSys)*, 2010.
7. A. Ambainis. Upper bound on the communication complexity of private information retrieval. In *Proc. International Colloquium on Automata, Languages and Programming*, pages 401–407. Springer-Verlag, 1997.
8. P. Bahl, R. Chandra, T. Moscibroda, R. Murty, and M. Welsh. White space networking with wi-fi like connectivity. In *Proc. ACM SIGCOMM Conference on Data communication (SIGCOMM)*, pages 27–38, 2009.
9. B. Barak, K. Chaudhuri, C. Dwork, S. Kale, F. McSherry, and K. Talwar. Privacy, accuracy, and consistency too: a holistic solution to contingency table release. In *Proc. ACM SIGMOD-SIGACT-SIGART symposium on Principles of database systems (PODS)*, 2007.
10. A. Beimel, Y. Ishai, E. Kushilevitz, and J.F. Raymond. Breaking the o (n< sup> 1</sup>(2k-1)/) barrier for information-theoretic private information retrieval. In *Foundations of Computer Science, 2002. Proceedings. The 43rd Annual IEEE Symposium on*, pages 261–270, 2002.
11. E. Beinat. Privacy and location-based: Stating the policies clearly. In *GeoInformatics*, pages 14–17, 2001.
12. A. Blum, K. Ligett, and A. Roth. A learning theory approach to non-interactive database privacy. In *Proc. ACM STOC*, 2008.
13. D. Cabric, S.M. Mishra, and R.W. Brodersen. Implementation issues in spectrum sensing for cognitive radios. In *Proc. IEEE Asilomar conf, Signals, Systems, and Computers*, volume 1, pages 772–776, 2004.
14. S. Chawla, C. Dwork, F. McSherry, A. Smith, and H. Wee. Toward privacy in public databases. In *Proc. International Conference on Theory of Cryptography*, pages 363–385. Springer-Verlag, 2005.

W. Wang and Q. Zhang, *Location Privacy Preservation in Cognitive Radio Networks*,
SpringerBriefs in Computer Science, DOI 10.1007/978-3-319-01943-7,
© The Author(s) 2014

15. D. Chen, S. Yin, Q. Zhang, M. Liu, and S. Li. Mining spectrum usage data: a large-scale spectrum measurement study. In *Proc. ACM International Conference on Mobile Computing and Networking (MobiCom)*, pages 13–24, 2009.

16. H. Chen and W. Gao. Btext on cyclostationary feature detector - information annex on sensing techniques. *IEEE 802.22 Meeting Doc.*, 2007.

17. R. Chen, N. Mohammed, B.C.M. Fung, B.C. Desai, and L. Xiong. Publishing set-valued data via differential privacy. *International Conference on Very Large Databases (VLDB)*, 4(11), 2011.

18. R. Chen, J.M. Park, and K. Bian. Robust distributed spectrum sensing in cognitive radio networks. In *Proc. IEEE International Conference on Computer Communications (INFOCOM)*, 2008.

19. R. Chen, J.M. Park, and J.H. Reed. Defense against primary user emulation attacks in cognitive radio networks. *IEEE J. Sel. Areas Commun.*, 26(1):25–37, 2008.

20. B. Chor, E. Kushilevitz, O. Goldreich, and M. Sudan. Private information retrieval. *Journal of the ACM (JACM)*, 45(6):965–981, 1998.

21. T.C. Clancy. Formalizing the interference temperature model. *Wireless Communications and Mobile Computing*, 7(9):1077–1086, 2007.

22. Federal Communications Commission. Spectrum policy task force report. *ET Docket No. 02-135*, 2002.

23. Federal Communications Commission. Establishment of interference temperature metric to quantify and manage interference and to expand available unlicensed operation in certain fixed mobile and satellite frequency bands. *ET Docket No. 03-289*, 2003.

24. Federal Communications Commission. Facilitating opportunities for flxible, efficient, and reliable spectrum use employing cognitive radio technologies. *ET Docket No. 03-108*, 2003.

25. Federal Communications Commission. Notice of proposed rule making and order. *ET Docket No. 03-322*, 2003.

26. Federal Communications Commission. Second report and order. *FCC 08-260*, Nov. 2008.

27. Federal Communications Commission. Second memorandum opinion and order. *FCC 10-174*, 2010.

28. Federal Communications Commission. Third memorandum opinion and order. *FCC 12-36*, May 2012.

29. C. Cordeiro, K. Challapali, D. Birru, and N. Sai Shankar. Ieee 802.22: the first worldwide wireless standard based on cognitive radios. In *Proc. IEEE Symposium on New Frontiers in Dynamic Spectrum (DySPAN)*, 2005.

30. G. Cormode, M. Procopiuc, E. Shen, D. Srivastava, and T. Yu. Differentially private spatial decompositions. In *Proc. IEEE International Conference on Data Engineering (ICDE)*, 2012.

31. F.F. Digham, M.S. Alouini, and M.K. Simon. On the energy detection of unknown signals over fading channels. *IEEE Trans. Commun.*, 55(1):21–24, 2007.

32. B. Ding, M. Winslett, J. Han, and Z. Li. Differentially private data cubes: optimizing noise sources and consistency. In *Proc. ACM SIGMOD International Conference on Management of Data (SIGMOD)*, 2011.

33. Wenliang Du and Zhijun Zhan. Using randomized response techniques for privacy-preserving data mining. In *Proc. ACM SIGKDD Conference on Knowledge Discovery and Data Mining (KDD)*, pages 505–510, 2003.

34. C. Dwork, F. McSherry, K. Nissim, and A. Smith. Calibrating noise to sensitivity in private data analysis. pages 265–284, 2006.

35. K. Efremenko. 3-query locally decodable codes of subexponential length. *SIAM Journal on Computing*, 41(6):1694–1703, 2012.

36. A. Evfimievski, J. Gehrke, and R. Srikant. Limiting privacy breaches in privacy preserving data mining. In *Proc. ACM SIGMOD-SIGACT-SIGART symposium on Principles of database systems (PODS)*, pages 211–222, 2003.

37. X. Feng, J. Zhang, and Q. Zhang. Database-assisted multi-ap network on tv white spaces: Architecture, spectrum allocation and ap discovery. In *Proc. IEEE New Frontiers in Dynamic Spectrum Access Networks (DySPAN)*, pages 265–276, 2011.

38. A. Friedman and A. Schuster. Data mining with differential privacy. In *Proc. ACM SIGKDD Conference on Knowledge Discovery and Data Mining (KDD)*, 2010.
39. Z. Gao, H. Zhu, Y. Liu, M. Li, and Z. Cao. Location privacy leaking from spectrum utilization information in database-driven cognitive radio network. In *Proceedings of the 2012 ACM conference on Computer and communications security*, Proc. ACM Computer and communications security, 2012.
40. Z. Gao, H. Zhu, Y. Liu, M. Li, and Z. Cao. Location privacy in database-driven cognitive radio networks: Attacks and countermeasures. In *Proc. IEEE International Conference on Computer Communications (INFOCOM)*, 2013.
41. W. Gasarch. A survey on private information retrieval. In *Bulletin of the EATCS*. Citeseer, 2004.
42. B. Gedik and L. Liu. Location privacy in mobile systems: A personalized anonymization model. In *Proc. IEEE International Conference on Distributed Computing Systems (ICDCS)*, 2005.
43. A. Ghasemi and E.S. Sousa. Opportunistic spectrum access in fading channels through collaborative sensing. *J. Commun.*, 2(2):71–82, 2007.
44. J. Girao, D. Westhoff, and M. Schneider. Cda: Concealed data aggregation for reverse multicast traffic in wireless sensor networks. In *Proc. IEEE International Conference on Communications (ICC)*, 2005.
45. M. Götz, A. Machanavajjhala, G. Wang, X. Xiao, and J. Gehrke. Publishing search logs- a comparative study of privacy guarantees. *IEEE Trans. on Knowl. and Data Eng.*, 2012.
46. M. Götz, S. Nath, and J. Gehrke. MaskIt: Privately releasing user context streams for personalized mobile applications. In *Proc. ACM SIGMOD International Conference on Management of Data (SIGMOD)*, pages 289–300, 2012.
47. M.M. Groat, W. He, and S. Forrest. KIPDA: k-indistinguishable privacy-preserving data aggregation in wireless sensor networks. In *Proc. IEEE International Conference on Computer Communications (INFOCOM)*, 2011.
48. M. Gruteser and D. Grunwald. Anonymous usage of location-based services through spatial and temporal cloaking. In *Proc. ACM International Conference on Mobile Systems, Applications and Services (MobiSys)*, pages 31–42, 2003.
49. M. Hay, V. Rastogi, G. Miklau, and D. Suciu. Boosting the accuracy of differentially private histograms through consistency. *International Conference on Very Large Databases (VLDB)*, 2010.
50. S. Haykin, D.J. Thomson, and J.H. Reed. Spectrum sensing for cognitive radio. *Proceedings of the IEEE*, 97(5):849–877, 2009.
51. B. Hoh, M. Gruteser, H. Xiong, and A. Alrabady. Enhancing security and privacy in traffic-monitoring systems. *IEEE Perv. Comput.*, 5(4):38–46, 2006.
52. S.Y. Huang. Intelligent decision support: handbook of applications and advances of the rough sets theory. Springer-Verlag New York, Inc., 1992.
53. J. Jia, Q. Zhang, and X. Shen. Hc-mac: A hardware-constrained cognitive mac for efficient spectrum management. *IEEE J. Sel. Areas Commun.*, 26(1):106–117, 2008.
54. J. Jia, Q. Zhang, Q. Zhang, and M. Liu. Revenue generation for truthful spectrum auction in dynamic spectrum access. In *Proc. ACM symposium on Mobile ad hoc networking and computing (MobiHoc)*, pages 3–12, 2009.
55. E. Kim, S. Helal, and D. Cook. Human activity recognition and pattern discovery. *IEEE Perv. Comput.*, 9(1):48–53, 2010.
56. V. Kindratenko, M. Pant, and D. Pointer. Deploying the olsr protocol on a network using sdr as the physical layer. Technical report, NCASSR Technical Report, UCLA, 2005.
57. M. Kodialam, T.V. Lakshman, and S. Mukherjee. Effective ad targeting with concealed profiles. In *Proc. IEEE International Conference on Computer Communications (INFOCOM)*, pages 2237–2245, 2012.
58. A. Korolova, K. Kenthapadi, N. Mishra, and A. Ntoulas. Releasing search queries and clicks privately. In *Proc. International Conference on World Wide Web (WWW)*, 2009.

59. L. Lai, Y. Fan, and H.V. Poor. Quickest detection in cognitive radio: A sequential change detection framework. In *Proc. IEEE Global Communications Conferences (Globecom)*, 2008.
60. K. Letaief and W. Zhang. Cooperative communications for cognitive radio networks. *Proceedings of the IEEE*, 97(5):878–893, 2009.
61. H. Li. Learning the spectrum via collaborative filtering in cognitive radio networks. In *Proc. IEEE Symposium on New Frontiers in Dynamic Spectrum (DySPAN)*, 2010.
62. H. Li and Z. Han. Catch me if you can: an abnormality detection approach for collaborative spectrum sensing in cognitive radio networks. *IEEE Trans. Wireless Commun.*, 9(11):3554–3565, 2010.
63. N. Li, T. Li, and S. Venkatasubramanian. t-closeness: Privacy beyond k-anonymity and l-diversity. In *Proc. IEEE International Conference on Data Engineering (ICDE)*, pages 106–115, 2007.
64. S. Li, H. Zhu, Z. Gao, X. Guan, K. Xing, and X. Shen. Location privacy preservation in collaborative spectrum sensing. In *Proc. IEEE International Conference on Computer Communications (INFOCOM)*, 2012.
65. B. Liu, Y. Jiang, F. Sha, and R. Govindan. Cloud-enabled privacy-preserving collaborative learning for mobile sensing. In *Proc. ACM Conference on Embedded Networked Sensor Systems (SenSys)*, 2012.
66. A. Machanavajjhala, D. Kifer, J. Gehrke, and M. Venkitasubramaniam. l-diversity: Privacy beyond k-anonymity. *ACM Trans. Knowl. Discov. Data*, 2007.
67. A. Mannini and A.M. Sabatini. Accelerometry-based classification of human activities using Markov modeling. *Intell. Neuroscience*, 2011(4):1–10, 2011.
68. Y. Matsuo, N. Okazaki, K. Izumi, Y. Nakamura, T. Nishimura, K. Hasida, and H. Nakashima. Inferring long-term user properties based on users' location history. In *Proc. International Joint Conference on Artificial Intelligence (IJCAI)*, pages 2159–2165, 2007.
69. F. McSherry and K. Talwar. Mechanism design via differential privacy. In *Proc. Annual Symposium on Foundations of Computer Science (FOCS)*, 2007.
70. F.D. McSherry. Privacy integrated queries: an extensible platform for privacy-preserving data analysis. In *Proc. ACM SIGMOD International Conference on Management of Data (SIGMOD)*, 2009.
71. Microsoft. Location based services usage and perceptions survey. 2011.
72. Microsoft. www.cs.cmu.edu/~CompThink/mindswaps/oct07/difpriv.ppt.
73. A.W. Min, K.G. Shin, and X. Hu. Secure cooperative sensing in ieee 802.22 wrans using shadow fading correlation. *IEEE Trans. Mobile Comput.*, 10(10):1434–1447, 2011.
74. A.W. Min, X. Zhang, and K.G. Shin. Detection of small-scale primary users in cognitive radio networks. *IEEE J. Sel. Areas Commun.*, 29(2):349–361, 2011.
75. N. Mohammed, R. Chen, B. Fung, and P.S. Yu. Differentially private data release for data mining. In *Proc. ACM SIGKDD Conference on Knowledge Discovery and Data Mining (KDD)*, 2011.
76. N. Mohammed, B. Fung, P.C.K. Hung, and C. Lee. Anonymizing healthcare data: a case study on the blood transfusion service. In *Proc. ACM SIGKDD Conference on Knowledge Discovery and Data Mining (KDD)*, 2009.
77. D. Niyato and E. Hossain. Competitive pricing for spectrum sharing in cognitive radio networks: Dynamic game, inefficiency of nash equilibrium, and collusion. *IEEE J. Sel. Areas Commun.*, 26(1):192–202, 2008.
78. R. Ostrovsky and W.E. Skeith. A survey of single-database private information retrieval: Techniques and applications. In *Proc. International Conference on Practice and Theory in Public-Key Cryptography*, pages 393–411. Springer-Verlag, 2007.
79. R. Paulet, M.G. Koasar, X. Yi, and E. Bertino. Privacy-preserving and content-protecting location based queries. In *Proc. IEEE International Conference on Data Engineering (ICDE)*, pages 44–53, 2012.
80. S. Peng, Y. Yang, Z. Zhang, M. Winslett, and Y. Yu. Dp-tree: indexing multi-dimensional data under differential privacy. In *Proc. ACM SIGMOD International Conference on Management of Data (SIGMOD)*, 2012.

81. B. Przydatek, D. Song, and A. Perrig. Sia: Secure information aggregation in sensor networks. In *Proc. ACM Conference on Embedded Networked Sensor Systems (SenSys)*, 2003.

82. Z. Quan, S. Cui, A.H. Sayed, and H.V. Poor. Optimal multiband joint detection for spectrum sensing in cognitive radio networks. *IEEE Trans. Signal Process.*, 57(3):1128–1140, 2009.

83. V. Rastogi and S. Nath. Differentially private aggregation of distributed time-series with transformation and encryption. In *Proc. ACM SIGMOD International Conference on Management of Data (SIGMOD)*, 2010.

84. V. Rastogi, D. Suciu, and S. Hong. The boundary between privacy and utility in data publishing. In *International Conference on Very Large Databases (VLDB)*, pages 531–542, 2007.

85. R. Rosales and G. Fung. Learning sparse metrics via linear programming. In *Proc. ACM SIGKDD Conference on Knowledge Discovery and Data Mining (KDD)*, 2006.

86. P. Samarati. Protecting respondents identities in microdata release. *IEEE Trans. on Knowl. and Data Eng.*, 2001.

87. S.J. Shellhammer, S.Sh. N, R. Tandra, and J. Tomcik. Performance of power detector sensors of DTV signals in IEEE 802.22 WRANs. In *Proc. ACM International Workshop on Technology and Policy for accessing Spectrum (TAPAS)*, 2006.

88. M. Shin, C. Cornelius, D. Peebles, A. Kapadia, D. Kotz, and N. Triandopoulos. Anonysense: A system for anonymous opportunistic sensing. *J. Perv. Mobile Comput.*, 7(1):16–30, 2011.

89. L. Sweeney. k-anonymity: A model for protecting privacy. *Int. J. Uncertain. Fuzziness Knowl.-Based Syst.*, 2002.

90. A. Taherpour, Y. Norouzi, M. Nasiri-Kenari, A. Jamshidi, and Z. Zeinalpour-Yazdi. Asymptotically optimum detection of primary user in cognitive radio networks. *Communications, IET*, 1(6):1138–1145, 2007.

91. K. Tan, H. Liu, J. Zhang, Y. Zhang, J. Fang, and G.M. Voelker. Sora: high-performance software radio using general-purpose multi-core processors. *Comm. ACM*, 54(1):99–107, 2011.

92. R. Tandra and A. Sahai. Snr walls for signal detection. *IEEE J. Sel. Topics in Signal Processing*, 2(1):4–17, 2008.

93. C. Tang. Obama administration calls for "privacy bill of rights". 2011.

94. Y. Tao, X. Xiao, J. Li, and D. Zhang. On anti-corruption privacy preserving publication. In *Proc. IEEE International Conference on Data Engineering (ICDE)*, pages 725–734, 2008.

95. A.G. Thakurta. Discovering frequent patterns in sensitive data. In *Proc. ACM SIGKDD Conference on Knowledge Discovery and Data Mining (KDD)*, 2010.

96. E. Toch, J. Cranshaw, P.H. Drielsma, J.Y. Tsai, P.G. Kelley, J. Springfield, L. Cranor, J. Hong, and N. Sadeh. Empirical models of privacy in location sharing. In *Proc. ACM International Conference on Ubiquitous Computing (Ubicomp)*, pages 129–138, 2010.

97. L. Trevisan. Some applications of coding theory in computational complexity. *arXiv preprint cs/0409044*, 2004.

98. J. Trostle and A. Parrish. Efficient computationally private information retrieval from anonymity or trapdoor groups. In *Proc. International Conference on Information Security*, pages 114–128. Springer-Verlag, 2011.

99. J. Vaidya and C. Clifton. Privacy preserving association rule mining in vertically partitioned data. In *Proc. ACM SIGKDD Conference on Knowledge Discovery and Data Mining (KDD)*, 2002.

100. J. Vaidya and C. Clifton. Privacy-preserving k-means clustering over vertically partitioned data. In *Proc. ACM SIGKDD Conference on Knowledge Discovery and Data Mining (KDD)*, 2003.

101. P.K. Varshney and C.S. Burrus. Distributed detection and data fusion. Springer-Verlag New York, Inc., 1997.

102. K. Vu, R. Zheng, and J. Gao. Efficient algorithms for k-anonymous location privacy in participatory sensing. In *Proc. IEEE International Conference on Computer Communications (INFOCOM)*, 2012.

103. A. Wald. Sequential tests of statistical hypotheses. *Ann. Math. Statist.*, 16(2):117–186, 1945.

104. Y. Wang, D. Xu, X. He, C. Zhang, F. Li, and B. Xu. L2P2: Location-aware location privacy protection for location-based services. In *Proc. IEEE International Conference on Computer Communications (INFOCOM)*, pages 1996–2004, 2012.

105. S.L. Warner. Randomized response: A survey technique for eliminating evasive answer bias. *Journal of the American Statistical Association*, 60(309):63–69, 1965.

106. S.B. Wicker. The loss of location privacy in the cellular age. *Commun. ACM*, 55(8):60–68, 2012.

107. R.C.W. Wong, J. Li, A.W.C. Fu, and K. Wang. (α,k) - anonymity: an enhanced k-anonymity model for privacy preserving data publishing. In *Proc. ACM SIGKDD Conference on Knowledge Discovery and Data Mining (KDD)*, 2006.

108. X. Xiao and Y. Tao. Personalized privacy preservation. In *Proc. ACM SIGMOD International Conference on Management of Data (SIGMOD)*, 2006.

109. X. Xiao, Y. Tao, and M. Chen. Optimal random perturbation at multiple privacy levels. *International Conference on Very Large Databases (VLDB)*, 2(1):814–825, 2009.

110. X. Xiao, G. Wang, and J. Gehrke. Differential privacy via wavelet transforms. In *Proc. IEEE International Conference on Data Engineering (ICDE)*, 2010.

111. J. Xu, Z. Zhang, X. Xiao, Y. Yang, and G. Yu. Differentially private histogram publication. In *Proc. IEEE International Conference on Data Engineering (ICDE)*, 2012.

112. S. Yekhanin. Towards 3-query locally decodable codes of subexponential length. *Journal of the ACM (JACM)*, 55(1):1, 2008.

113. S. Yekhanin. Locally decodable codes and private information retrieval schemes. Springer-Verlag New York, Inc., 2010.

114. S. Yekhanin. Private information retrieval. *Communications of the ACM*, 53(4):68–73, 2010.

115. H. Yu, X. Jiang, and J. Vaidya. Privacy-preserving svm using nonlinear kernels on horizontally partitioned data. In *Proc. ACM Symposium on Applied Computing (SAC)*, 2006.

116. S. Yu, G. Fung, R. Rosales, S. Krishnan, R. B. Rao, C. Dehing-Oberije, and P. Lambin. Privacy-preserving cox regression for survival analysis. In *Proc. ACM SIGKDD Conference on Knowledge Discovery and Data Mining (KDD)*, 2008.

117. Q. Zhang, J. Jia, and J. Zhang. Cooperative relay to improve diversity in cognitive radio networks. *Communications Magazine, IEEE*, 47(2):111–117, 2009.